ISBN 978-0-282-82128-9
PIBN 10866803

This book is a reproduction of an important historical work. Forgotten Books uses
state-of-the-art technology to digitally reconstruct the work, preserving the original format
whilst repairing imperfections present in the aged copy. In rare cases, an imperfection in
the original, such as a blemish or missing page, may be replicated in our edition. We do,
however, repair the vast majority of imperfections successfully; any imperfections that
remain are intentionally left to preserve the state of such historical works.

1 MONTH OF
FREE
READING

at

www.ForgottenBooks.com

English
Français
Deutsche
Italiano
Español
Português

www.forgottenbooks.com

Mythology Photography **Fiction**
Fishing Christianity **Art** Cooking
Essays Buddhism Freemasonry
Medicine **Biology** Music **Ancient
Egypt** Evolution Carpentry Physics
Dance Geology **Mathematics** Fitness
Shakespeare **Folklore** Yoga Marketing
Confidence Immortality Biographies
Poetry **Psychology** Witchcraft
Electronics Chemistry History **Law**
Accounting **Philosophy** Anthropology
Alchemy Drama Quantum Mechanics
Atheism Sexual Health **Ancient History**
Entrepreneurship Languages Sport
Paleontology Needlework Islam
Metaphysics Investment Archaeology
Parenting Statistics Criminology
Motivational

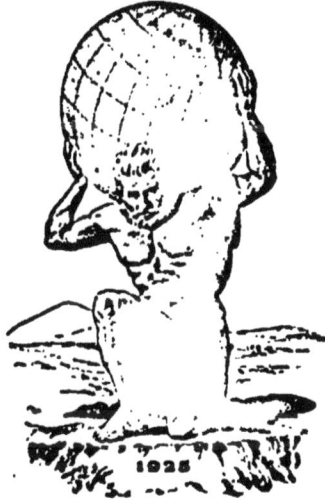

THE

COMPLETE

PRACTICAL BREWER;

OR,

PLAIN, ACCURATE, AND THOROUGH INSTRUCTIONS

IN THE

ART OF BREWING ALE, BEER, AND PORTER;

INCLUDING THE PROCESS OF MAKING

BAVARIAN BEER; ALSO, ALL THE SMALL BEERS,

SUCH AS

ROOT-BEER, GINGER-POP, SARSAPARILLA-BEER, MEAD, SPRUCE-BEER, ETC. ETC.

ADAPTED TO

THE USE OF PUBLIC BREWERS,

AND

PRIVATE FAMILIES, OR THOSE WHO MAY WISH TO BREW ON A SMALL SCALE.

With Numerous Illustrations.

By M. L. BYRN, M.D.

GRADUATE OF THE UNIVERSITY OF THE CITY OF NEW YORK; AUTHOR OF "DETECTION OF FRAUD AND PROTECTION OF HEALTH," ETC. ETC.

HENRY CAREY BAIRD,

SUCCESSOR TO E. L. CAREY.

NO. 7 HART'S BUILDINGS, SIXTH ST. ABOVE CHESTNUT.

1852.

Entered according to Act of Congress, in the year 1852, by
HENRY CAREY BAIRD,
in the Clerk's Office of the District Court for the Eastern District of
Pennsylvania.

STEREOTYPED BY L. JOHNSON & CO.
PHILADELPHIA.

PRINTED BY T. K. & P. G. COLLINS.

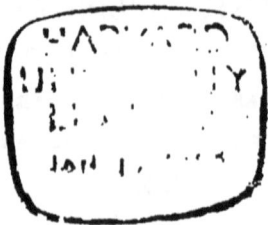

C. W. Barksdale's &c.
Book
Sept 20th 1856

C W Barksdale

C W Barksdale
Book
in Sept 20th 185

PREFACE.

In offering this work to the public, I do not lay claim to originality, or profess to be giving information that is in itself entirely new, as a reason for preparing a book on the subject; but I claim that I have prepared a book which has long been called for, and which has never been obtained before— "A COMPLETE PRACTICAL BREWER." Much valuable information on the subject of Brewing has been imparted in large books, such as Encyclopedias, Dictionaries, etc., and one or two large works exclusively on this subject have been written in Europe; but the great fault in all those is that they are too voluminous; and of course, being of such size, they were sold at a corresponding price, which, in

many instances, was objectionable. Besides the
price, there was a still greater objection—too much
matter had to be looked over to obtain the plain,
simple processes of brewing; and those who were
seeking information on the subject, would become
fatigued and worn out before perusing such a large
volume. And even when carefully studied, there
was so much idle speculation intermingled with useful facts, that it was almost impossible to arrive at
correct conclusions. In the present volume, every-
thing that is practical and useful has been given,
and all idle speculation dispensed with which could
not benefit the brewer in his operations of making
Beer, Porter, &c.

The process of making all kinds of Small Beers,
which has not been spoken of before in any scientific
work on Brewing, is here given in detail, and must
prove of great practical utility.

Every thing has been so simplified that, with any
ordinary ingenuity, almost every person may erect
a brewery and put it in successful operation. The
assistance of a mechanic will, of course, be very

beneficial, and one should always be employed when practicable.

. It is almost needless to say that every authority that is worthy of notice has been consulted in getting up this work, the enumeration of which, to the reader, would be useless, and they are therefore not given.

M. L. B.

New York, *September 1st, 1852.*

CONTENTS.

THE

COMPLETE PRACTICAL BREWER.

GRAIN USED FOR BREWING.

THE grain generally made use of in brewing, and which answers the purpose best, is barley. Besides this, many other grains are used for the purpose of making beer, in greater or less quantity. The common Indian corn is often used in this country for making beer, but not in any other that I know of. The process of converting it into malt is not that which has been stated by some British authors, and others, viz. " burying the grain under the ground ; and when germination has made sufficient progress, it is dug up and kiln-dried." The process consists in putting a quantity of the grain into a large hogshead, or other suitable vessel, with perforations in the bottom for the water to escape, and keeping it moistened with warm water until germination has commenced ; it is then left for the germinating process to proceed far enough, when it is taken out and dried in the usual manner. Some persons may " *bury*" the grain in this country, as stated by some writers ; but if so, I have never known of it. But even if they did, it would not be as good a method as the one I have spoken of, for the grain would necessarily partake of the *earthy taste,* and render it unfit for making a pleasant beverage.

As regards other substances used in brewing, I will speak of them under another head. Barley is the seed of the *Hordeum vulgare,* a plant which has long been cultivated principally for the fabrication of beer. Of the *Hordeum,* there are two species cultivated, both in this country and Europe. One is the *Hordeum vulgare,* or barley in which the seeds are disposed in two rows on the spike; the other is the *Hordeum hexastichon,* called frequently *bigg.* We observe in this species that the grains are disposed in two rows, as in the other; but three seeds spring from the same point, so that the head of *bigg* appears to have the seeds disposed in six rows. The *bigg* is a much more hardy plant than barley, and ripens more rapidly. It is for this reason that it thrives better than barley in high and cold situations. The genuine barley is desirable when you wish to make the best beer. The grains of the bigg are not so large as those of barley.

MALTING.

BARLEY is usually converted into malt before employing it in the manufacture of ale; but it is not absolutely necessary, as has been proved by numerous experiments, made by different individuals, though it greatly facilitates the process. In making ale from unmalted barley, several precautions are necessary in order to succeed. For instance, the water let upon the ground barley in the mash-tun must be considerably below the boiling temperature, for barley meal is much more apt to *set* than malt—that is, to form a stiff paste, from which no wort will separate. But the addition of a portion of the chaff of oats serves considerably to prevent this *setting of the goods*, and facilitates very much the separation of the wort.

Great care must be taken, likewise, to prevent the heat from escaping during the mashing; and the mashing should be continued longer than usual; for it is during the mashing that the starch of the barley is converted into saccharine matter. The change spoken of seems to be owing simply to the chemical combination of a portion of water with the starch of the barley; precisely as happens when common starch is converted into sugar by boiling with very dilute sulphuric or any other acid. This method of brewing unmalted barley answers admirably for small beer. It is thought, by many, that the raw barley does not answer for making strong ale; and the beer made from it is said to have a peculiarly unpleasant taste, though it keeps for years without running into acidity.

Malting consists of the following processes:—*Steeping, couching, flooring, sweating,* and *kiln-drying.*

The *steeping* is performed in large cisterns, made of wood or stone, which being filled with clear water up to a certain height, a quantity of barley is shot into them and well stirred about with rakes. Grain that is good is heavy, and subsides; the lighter grains that float on the surface are damaged, and are to be skimmed off; for they would damage the quality of the malt and the flavour of the beer made with it. Seldom do they amount to more than two per cent. Portions of barley are successively emptied into the steep cistern, till the water stands only a few inches (about five) above its surface; when this is levelled very carefully, and every light seed is removed.

Generally, the steep lasts from forty to sixty hours, varying, however, according to circumstances: new barley requires a longer period than old, and bigg requiring much less time than barley.

In the course of this steep, some carbonic acid is evolved from the grains and combines with the water, which, at the same time, takes on a yellowish tinge, and acquires a smell resembling straw, from the fact that it dissolves some of the extractive matter of the barley-husks. The grain imbibes about one-half of its weight of water, and increases in size about one-fifth. By losing this extract the husk becomes nearly one-seventieth lighter in weight, and paler in colour.

The length of time that the grain continues in steep depends, in a slight measure, on the temperature of the air, and is not so long in summer as in winter. Usually, from forty to forty-eight hours will be found sufficient for sound, dry grain. The object of steeping is to expand the farina of the barley with humidity, and thus prepare the seed for germination, in the same way as the moisture of the earth prepares for the growth of the radicle and plumule, in seed sown in it. The grain must not remain too long in the steep; it is injurious, because it prevents the germination at the proper time, and thus exhausts a

.portion of the vegetative power: it causes, also, an abstraction of saccharine matter by the water. Maceration is known to be complete when the grain can be easily transfixed with a needle, and is swollen to its full size. Here is what is thought to be a good test:—Should a barley-corn, when pressed between the finger and thumb, continue entire in its husk, it is not sufficiently steeped; but if it sheds its flour upon the finger, it is ready. Should the substance exude in the form of a milky juice, the steep has been too long continued, and the barley is spoiled for germination.

It sometimes happens, in warm weather, that the water becomes acid before the grain is thoroughly swelled. The way to avoid this accident, which is generally very evident to the smell and taste, is by drawing off the foul water through the tap at the bottom of the cistern, and replacing it with fresh cold water. It is well to remove the water two or three times at one steep.

It cannot be denied that carbonic acid is evolved during the steeping of grain. It is obvious from the most simple experiment. You have only to mix the steep-water with lime-water, and the whole becomes milky, and carbonate of lime is deposited. If the steep-water be agitated, it froths on the surface like ale.

After the water is drawn off, and occasionally a fresh quantity passed through, to wash away any slimy matter, which is apt to generate in warm weather, the barley is laid on the couch-floor, of stone flags or movable wooden boards, as the case may be, in heaps from twelve to sixteen inches high, and left in that position for twenty-four hours. When you can take the barley between your.

finger and thumb and squeeze it together, it has been long enough in the steep.

Should a thermometer be plunged into the grain and observed from time to time, it will be found that the barley continues for several hours without acquiring any perceptible increase of heat. The moisture on the surface of the corns during this period gradually exhales, or is absorbed, so that they do not perceptibly moisten the hand. At last, the thermometer begins to rise, and continues to do so gradually till the temperature of the grain is nearly ten degrees higher than that of the surrounding atmosphere. It is about ninety-six hours after it has been thrown out of the steep that this happens. An agreeable odour is now exhaled, which has some resemblance to that of apples. Should the hand be now thrust into the heap, it will be found that it feels warm, while, at the same time, it has become so moist as to wet the hand. This moisture, when it appears, is called *sweating*, by maltsters, and it constitutes a remarkable period in the process of malting. It is thought that a little alcohol is at this period exhaled by the grain.

If the grains in the inside of the heap at the time of sweating be examined, it will be observed that the roots are beginning to make their appearance at the bottom of each seed. They at first have the appearance of a white prominence, which soon divides itself into three rootlets. The number of rootlets in bigg seldom exceeds three, but in barley it frequently amounts to five or six. Unless their growth is checked, these rootlets increase in length with great rapidity; and the principal attention of the maltster is directed to keeping them short till the grain be sufficiently malted.

The very rapid growth of the roots, and the too high elevation of temperature, is prevented by spreading the grain thinner upon the floor, and carefully turning it over several times a day. The depth, at first, is about sixteen inches; but this depth is diminished a little at every turning, till at last it is reduced to three or four inches.

The turnings are to be regulated by the temperature of the malt, but they are seldom fewer in number than two each day. The temperature of the grain is kept as nearly as possible at fifty-five degrees, in Scotland; but in England, the temperature is about sixty-two degrees. .

In about twenty-four hours after the sprouting of the roots, the rudiment of the future stem begins to make its appearance. The name given to this substance is the *acrospire*. From the same extremity of the seed with the root this rises, and, advancing within the husk or skin, would at last (if the process were continued long enough) issue from the other extremity in the form of a green leaf; but the process of malting is stopped before the acrospire has made such progress.

During the time that the grain is on the malt-floor, it has been ascertained that it absorbs oxygen gas and throws off carbonic acid gas; but to what amount these absorptions and emissions take place has not been ascertained: they are certainly small. The appearance of the kernel, or mealy part of the corn, undergoes a considerable change as the acrospire shoots along the grain. The glutinous and mucilaginous matter in a great measure disappears, the colour becomes whiter, and the texture of the grain so loose that it crumbles to powder between the fingers. As soon as this is accomplished, which takes place when the acrospire has come nearly to the end of the seed, the process is stopped altogether.

It was formerly the custom in many breweries, at this period to pile up the whole grain into a pretty thick heap, and allow it to remain for some time. The evolution of a very considerable heat is the consequence, while, at the same time, the malt becomes exceedingly sweet. This plan, though, is now laid aside, because it occasions a sensible diminution in the malt, without being of any essential service; for the very same change takes place afterward, while the malt is in the mash-tun, without any loss whatever.

The length of time during which the grain continues on the malt-floor varies according to circumstances. It is converted into malt more speedily the higher the temperature of the grain is kept. Generally, fourteen days may be specified as the period which intervenes from throwing the barley out of the steep till it is ready for the kiln; though in some countries, Scotland for one, the time is not shorter than eighteen days, or even three weeks. Here, no doubt, is an advantage which one malting possesses over another, as every thing which shortens the progress, without injuring the malt, must turn out to the advantage of the manufacturer.

In very dry weather it is sometimes necessary to water the barley on the couch. Occasionally, the odour disengaged from the couch is offensive, resembling that of rotten apples. This is a bad prognostic, indicating either that the barley was of bad quality, or that the workmen, through careless shovelling, have crushed a number of the grains in turning them over.

For this reason, when the weather causes too quick germination, it is better to check it by spreading the heap out thinner than by turning it too frequently over.

A moderate temperature of the air is best adapted to malting: therefore, it cannot be carried on well during the heat of summer or the extreme cold of winter. Malt-floors should be placed in substantial thick-walled buildings, without access of the sun, so that a uniform temperature of 59° or 60° may prevail inside. Some recommend them to be sunk a little under the surface of the ground, if the situation is dry.

To dry the malt upon the kiln is the last process in malting, which stops the germination, and enables the brewer to keep the malt for some time without injury. As soon as the malt has become perfectly dry to the hand upon the floor, it is taken to the kiln, and dried hard with artificial heat, to stop all further growth. The malt-kiln, which is described particularly hereafter, is a round or square chamber, covered with perforated plates of cast-iron, whose area is heated by a stove or furnace, so that not merely the plates on which the malt is laid are warmed, but the air which passes up through the stratum of malt itself has the effect of carrying off very rapidly the moisture from the grains. The layer of malt should be about three or four inches thick, and evenly spread, and its heat should be steadily kept at from 90° to 100° Fahrenheit, till the moisture be mostly exhaled from it. During this time the malt must be turned over frequently, and latterly every three or four hours.

2*

When it is nearly dry, its temperature should be raised to from 145° to 165° F., and it must be kept at this heat till it has assumed the desired shade of colour, which is commonly a brownish-yellow, or a yellowish-brown. The fire is now allowed to die out, and the malt is left on the plates until it has become completely cool,—a result promoted by the stream of cool air which now rises up through the bars of the grate; or the thoroughly dry, browned malt may, by damping the fire, be taken hot from the plates, and cooled upon the floor of an adjoining appartment. The prepared malt must be kept in a dry loft, where it can be occasionally turned over until it is used. The period of kiln-drying should *not be hurried*. Many persons employ two days in this operation.

According to colour and degree of drying, malt is distributed into three sorts—pale, yellow, and brown. The first is produced when the highest heat to which it has been subjected is from 90° to 100° F.; the amber-yellow when it has suffered a heat of 122°; and the brown when it has been treated as above described. The *black malt* used by the porter brewer to colour his beer has suffered a much higher heat, and is partially charred. The temperature of the kiln should, in all cases, be most gradually raised, and most equably maintained. If the heat be too great at the beginning, the husk gets hard and dried, and hinders the evaporation of the water from the interior substance; and should the interior be dried by a stronger heat, the husk will probably split, and the farina become of a horny texture, very refractory in the mash-tun.

In general, it is preferable to brown malt rather by a

long-continued, moderate heat, than by a more violent
heat of shorter duration, which is apt to carbonize a portion of the mucilaginous sugar, and to damage the article.
In this way, the sweet is sometimes converted into a bitter
principle.

Good malt is distinguished by the following characters:
The grain is round and full, breaks freely between the
teeth, and has a sweetish taste, an agreeable smell, and is
full of a soft flour from end to end. It affords no unpleasant flavour on being chewed: it is not hard, so that when
drawn along an oaken table across the fibres, it leaves a
white streak, like chalk. It swims upon water, while unmalted barley sinks in it. Since the quality of the malt
depends much on that of the barley, the same sort only
should be used for one malting. New barley germinates
quicker than old, which is more dried up: a couch of a
mixture of the two would be irregular, and difficult to
regulate.

Description of the Malt-kiln.—Figs. 1, 2, 3, 4 show
the construction of a well-contrived *malt-kiln.* Fig. 1 is

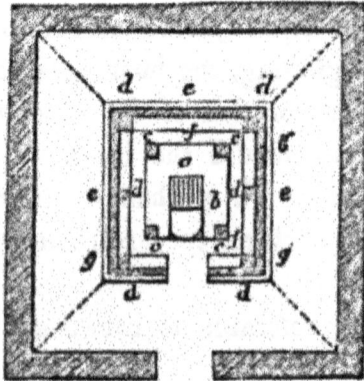

Fig. 1.

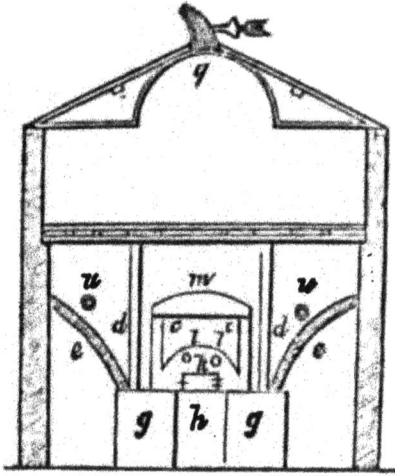

Fig. 2.

the ground-plan; fig. 2 is the vertical section; and figs.
3 and 4 a horizontal and vertical section in the line of
the malt-plates. The same letters denote the same parts
in each of the figures. A cast-iron cupola-shaped oven is
supported in the middle upon a wall of brickwork four
feet high; and beneath it are the grate and its ash-pit.

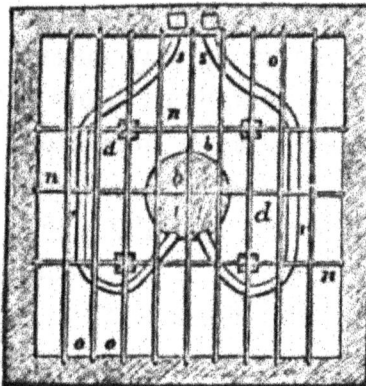

Fig. 3.

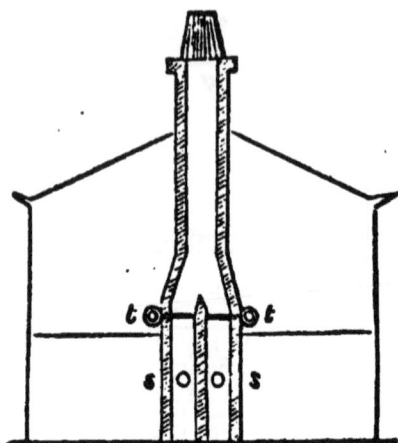

Fig. 4.

The smoke passes off through two equidistant pipes into the chimney. The oven is surrounded with four pillars, on whose top a stone lintel is laid: *a* is the grate, nine inches below the sole of the oven *b*; *c, c, c, c* are the four nine-inch strong pillars of brickwork which bear the lintel *m*; *d, d, d, d* are strong nine-inch pillars, which support the girder and joists upon which perforated plates repose; *e* denotes a vaulted arch on each of the four sides of the oven; *f* is the space between the kiln and the side-arch, into which a workman may enter to inspect and clean the kiln; *g, g,* the walls on either side of the kiln, upon which the arches rest; *h,* the space for the ashes to fall; *k,* the fire-door of the kiln; *l, l,* junction-pieces to connect the pipes *r, r* with the kiln: the mode of attaching them is shown in fig. 3. These smoke-pipes lie about three feet under the iron plates, and at the same distance from the side-walls: they are supported upon iron props, which are made fast to the arches. In fig. 2, *u* shows their section; at *s, s,* fig. 3, they enter the chimney, which is provided with two register or damper plates, to regulate the draught through the pipes. These registers are represented by *t, t,* fig. 4, which shows a perpendicular section of the chimney. *m,* fig. 2, is the lintel, which causes the heated air to spread laterally instead of ascending in one mass in the middle, and prevents any combustible particles from falling upon the iron cupola. *n, n* are the main girders of iron for the iron beams *o, o,* upon which the perforated plates *p* lie; *q,* fig. 2, is the vapour-pipe in the middle of the roof, which allows the steam of the drying malt to escape. The kiln may be heated either with coal or wood.

The size of this kiln is about twenty feet square; but it may be made proportionally either smaller or greater. The perforated floor should be large enough to receive the contents of one steep or couch.

The perforated plate might be conveniently heated by steam-pipes, laid zigzag, or in parallel lines under it; or a wire-gauze web might be stretched upon such pipes. The wooden joists of a common floor would answer perfectly to support this steam-range, and the heat of the pipes would cause an abundant circulation of air. For drying the pale malt of the ale brewer, this plan is particularly well adapted.

The kiln-dried malt is sometimes ground between stones in a common corn-mill, like oatmeal; but it is more generally crushed between iron rollers, in England.

The Crushing-mill.—The cylinder malt-mill is constructed as shown in figs. 5, 6. I is the sloping trough, by which the malt is let down from its bin or floor to the hopper A of the mill, whence it is progressively shaken

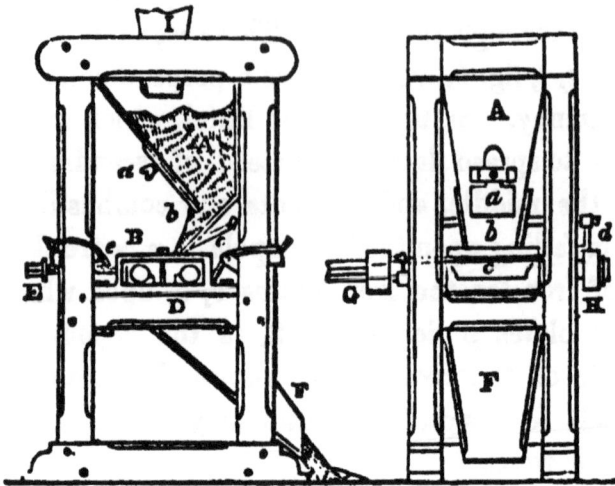

Fig. 5. Fig. 6.

in between the rollers B, D. The rollers are of iron, truly cylindrical, and their ends rest in bearers of hard brass, fitted into the side-frames of iron. A screw E goes through the upright, and serves to force the bearer of the one roller toward that of the other, so as to bring them closer together when the crushing effect is to be increased. G is the square end of the axis, by which one of the rollers may be turned either by the hand or by power: the other derives its rotatory motion from a pair of equal-toothed wheels H, which are fitted to the other end of the axes of the rollers. *d* is a catch which works into the teeth of a ratchet-wheel on the end of one of the rollers (not shown in this view.) The lever *c* strikes the trough *b* at the bottom of the hopper, and gives it the shaking motion for discharging the malt between the rollers, from the slide-sluice *a*. *e, e*, fig. 5, are scraper-plates of sheet-iron, the edges of which press by a weight against the surfaces of the rollers, and keep them clean.

Instead of the cylinders, some employ a crushing-mill of a conical, grooved form, like a coffee-mill upon a large scale.

It is probable that more of the kernel would be dissolved if the malt were ground finer than is customary to do. The reason for grinding it only coarsely, is to render it less apt to *set*. But this object might be accomplished equally well by bruising the malt between rollers, which would reduce the starchy part to powder, without destroying the husk. This method, indeed, is practised by many brewers, but it should be followed by *all*.

BREWING.

THIS consists of five successive processes, and they have been designated as follows:—MASHING, BOILING, COOLING, FERMENTING, CLEANSING.

Suppose, for the sake of stating the comparative quantities, that the object is to employ, in a single brewing, fifty bushels of malt. The first thing to be done is to grind the malt in a mill; and the best kind of mill for the purpose is that in which the malt is made to pass between two iron rollers.

A copper boiler must be provided, capable of boiling at least fifty bushels of malt; or its solid contents must, at the smallest, amount to 382 ale-gallons, which are 62½ cubic feet. It is necessary to place this copper boiler over brickwork, upon a furnace, and there must be conveniences for filling it with water, and for letting the water off, when sufficiently heated, into the mash-tun.

The mash-tun is a wooden vessel, composed of staves, properly fixed by means of iron hoops, and commonly placed in the middle of the brew-house. There is a false bottom to it, full of holes, at some little height above the true bottom. Its size varies, according to the extent of the brewing establishment; but a mash-tun capable of mashing fifty bushels of malt must be at least one-third larger than the bulk of malt, or it must be capable at least of containing seventy-five bushels.

To brew twenty barrels of ale, the open boiler must be capable of holding thirty-five barrels of wort; or, in other words, the boiler should contain a quantity of water about two-thirds more than the quantity of finished ale required. When an upperback is used, or when a condensing-pan is placed on the top, less room is required for the wort; but it is always better for the brewer to have the boiler above the standard than under it.

In mashing, there are two methods : one is by letting the water rise up through the malt from the false bottom of the mash-tun—the first mash at 170° heat; the second at 190°. The other method is by first filling into the mash-tun the whole quantity of water for the first mash at 180° of heat, and running the malt into it from the hopper above, stirring, at the same time, either with oars or by the machine. The second mash at 190°.

In both methods, the temperature of the water in the

first mash is lowered about 40°; but the mash afterward rises 20°, from the chemical action of the malt upon the water. You should regulate the quantity of water for the first mash by the required strength of the wort. It takes time for the flowing of the wort, according to the quantity in operation. If the mash-tun is large, the stop-cock is made large in proportion to the size of the mash-tun, making allowance for the velocity with which the worts escape by the pressure. It will require half an hour for the flowing of a brewing of thirty barrels; though when sparging is taken into account, it may take six hours to finish the mashing and flowing of the wort.

When the water is mixed with the malt, the mixture is completely stirred and all the clots broken, either by workmen, who use for the purpose very narrow wooden shovels, or, when the capacity of the mash-tun is very great, as in the London breweries, by a machine which is driven by a steam-engine. Particular care must be taken to break all the clots, because the whole of the malt within them would otherwise escape the action of the water and be lost to the brewer.

So soon as the water and malt are sufficiently mixed, the mash-tun is covered, and left in this state about three hours. The time varies, though, according to circum-stances.

The specific gravity of water is less than malt-corn: still, the corn will swim on the surface of water. It is accounted for from the fact that there is lodged between the skin and the kernel a quantity of air, which it is not easy to drive away. Thus we see that brewers are in the habit of judging of the goodness of malt by throwing a certain quantity of it into the water, and reckoning the grains which fall to the bottom : these indicate the proportion of unmalted grain which the malt contains. Of course, the more of them that exist in any given quantity of malt the worse the malt must be considered. But though malt, when we consider only single corns, is about one-sixth heavier than water, yet a bushel of malt does not weigh so much as one-third of a bushel of water.

When the mash has continued for three hours, (longer or shorter, according to circumstances,) a stopcock, placed below the false bottom in the mash-tun, is opened, and the wort allowed to run out into a vessel prepared to receive it, and known by the name of *underback*. Also, at the same time, the cover is taken off the mash-tun, and quantities of water of the temperature of 180° are occasionally sprinkled over it from the boiler, which had been again filled with water to be heated as soon as the water for mashing was drawn off. Specific directions cannot be given respecting the quantity of hot water added in this manner by sprinkling, because that must depend upon the views of the brewer. Should he wish to have ale of very great strength, he will of course add less water: if the ale is to be weak, he will add more. A good plan is to determine the strength of the liquor as it flows from the underback, by means of a saccharometer, or by taking its specific gravity. If the specific gravity (at 60°) sinks to 1.04 or 1.05, or if it contains only 36½ or 46½ lbs. per barrel of solid matter in solution, it would be useless or injurious to draw any more off for making strong ale. But an additional portion may still be drawn off and converted into small-beer.

Some fifty or sixty years ago, it was customary with some of the small-beer brewers in Edinburgh to make the small-beer of considerable strength; and after the exciseman had determined its quantity and the duty to be paid on it, they diluted it largely with water just when they were sending it out of the house. It was very easy to put this fraud in practice, because the small-beer is usually disposed of the moment it is mixed with the yeast, and before it has undergone any fermentation whatever. Fermentation goes on sufficiently in the small casks in which it is sent to the consumers. In many places it is customary to bottle this small-beer, which makes it clear and very brisk, and consequently very agreeable to the palate.

. No general rule can be laid down either for the specific gravity or strength when it begins to flow from the mash. Obviously, it will depend upon the goodness of the malt, and upon the quantity of mashing-water employed, when compared with the quantity of malt. When the wort first flows from the mash-tun, it is a transparent liquid, of a fine amber colour, a peculiar smell, and a rich, luscious, sweet taste. Should it show cloudiness, as sometimes happens, it is a proof that the water used for mashing was of too high a temperature.

It requires at least six or eight hours for the flowing of the wort from the mash-tun. During its progress the colour diminishes, the smell becomes less agreeable, and the taste less sweet. Finally the colour becomes nearly opal, the smell becomes sour, and somewhat similar to the odour emitted by an infusion of meal and water left until it has become sour; still it produces no change on vegetable blue colours.

According to the experiments of Saussure, it would appear that starch-sugar is nothing else than a combination of starch and water. For this reason it is probable, that during the mashing, a combination takes place between the starch of the malt and the water, the result of which is the formation of starch-sugar. The properties of this sugar agrees very much with the sugar of grapes. It crystallizes in needles grouped together in the form of small sphericles like granulated honey. Its sweetening power does not come up to that of the common sugar, and, like sugar of grapes, it ferments without the addition of yeast. Attempts have in vain been made to separate the saccharine part of the residue of wort from the starch. If alcohol is poured over it, no solution takes place; but such is the affinity of the residue of wort for water, that it deprives the alcohol of a portion of its water, just as carbonate of potash or muriate of lime does, and a very viscid liquid, consisting of the residue of malt dissolved in a very small quantity of water, is formed at the bottom of the vessel.

There is great difficulty experienced in evaporating wort without partly decomposing the extractive residue. The usual plan is to put it upon a very flat dish, and to apply a heat not greater than 120°.

There is very little saccharine matter contained in the wort that runs off last; but some starch and mucilaginous matter may still be detected in it. The beauty and flavour of the ale is increased if the wort only is taken that runs off first, and throw away the last drawn worts, or employ them only in the manufacture of small beer. Many brewers, though, differ in practice when drawing off wort. Should the' whole be intended for ale, the first mash is laid on at a much greater length, to obtain the greatest possible quantity of wort of a required strength. There is in the remaining mash, wort of the same strength as that drawn off. The second mash not only takes up the worts which saturate the goods, but the formation of saccharum still proceeds; and when this mash is run, supposing its weight 40 or 45 pounds per barrel, the saccharum that remains in the mash is only that contained in the wort which is taken up by the malt left in the mash-tun. The water run into the mash, say ten barrels, added to that which saturates the malt, will make fifteen barrels, of the weight of 13 pounds of saccharum per barrel, allowing wort of the weight of 40 pounds to have been left in the second mash. Many brewers, consequently, use the third mash for small-beer, as, were they to mix this weak third mash with the two first, they would lose more by boiling down to strength than its worth, besides damaging their ale.

The weight of the saccharine extract of the first and second mash of a brewing of ale will be in proportion to the required price of production.

The loss of the wort by evaporation on the coolers is so very great, that it is rendered an object of momentous consideration to devise some plan of cooling without so much loss. By running the wort through pipes of great length, immersed in water, distillers, who run their wort into coolers from the mash, have accomplished their object as completely as can be. This method of cooling, though, does not answer the brewer of ales, owing to the fecula remaining in solution and damaging the quality of the production, when such a plan is adopted.

The coolers which answer the purpose better than any other are those made of iron-plate, and are certainly preferable to the ordinary wooden ones now in use. In the iron cooler, by lowering its temperature by running cold water over it, and mopping it clean and dry, wort, by being then spread to the depth of 1½ or 2 inches successively, may be cooled down, even in summer-time, to as low a degree as brewers require.

Boiling of the worts is the next process in brewing. The wort is pumped up from the underback into the copper boiler, where it is boiled till it has acquired the degree of strength which is wanted by the brewer.

There is a flocky precipitate formed during the boiling of the wort, which, as far as has been ascertained, approaches nearly to the nature of gluten or vegetable albumen, for these two substances differ very little from each other. When the wort is in the boiler, the requisite quantity of *hops* is added to flavour the ale and render it capable of being kept for a considerable length of time without souring. As is well known, hops are the seed-pots of the *Humulus lupulus,* or *hop-plant,* which is cultivated in considerable quantities in the South of England, and also in the United States. The seed-pots of this creeping plant are collected when ripe, and dried upon a kiln. They are then packed up in bags, and sold to the brewer.

It is well known that hops have a peculiar bitter taste and a weak aromatic odour, and they possess narcotic or sedative qualities to a considerable extent. A pillow filled with hops has often been found to induce sleep after every other remedy had failed.   When they are digested for several days in alcohol, that liquid acquires a slight greenish colour, a peculiar taste, and an odour in which that of the hop can be distinctly observed.   Should the alcohol, previously freed from the undissolved matter, be distilled in a retort, there would remain behind a solid green-coloured oil.   Hops owe their peculiar flavour to this oil.   It has a taste peculiar, sharp, and scarcely bitter, but putting one in mind of the peculiar flavour of "good ale." This oil is the part of the hops which gives ale its peculiar flavour.   By long boiling it is very liable to be dissipated.   Thus we see the necessity, when hops are too long boiled in wort, that the aromatic odour and peculiar flavour should be nearly dissipated, and a bitter taste substituted.

Brewers are of the opinion that the intoxicating qualities of ale are to be partly ascribed to the oil of the hops. It has been pretty common, in fact, to ascribe intoxicating qualities to bitter-tasted substances in general; but this opinion, though general, does not appear to be founded upon any precise experiments or observations.   There is no volatile oil now known, I believe, that produces intoxication; though some of them, as oil of turpentine, act with great energy on the stomach.   No infusion of any bitter whatever, not even of hops, is known to produce intoxication; nor is any effect in the least similar to intoxication produced when considerable quantities (two or three ounces per day) of Peruvian bark are swallowed in substance.

Independent of the volatile oil, hops contain likewise a quantity of bitter principle, which can be easily extracted from them by water. So far as is now known, this bitter matter possesses the characters of the bitter principle in perfection. There is no reagent capable of throwing it down except acetate of lead—a somewhat doubtful precipitate, because it throws down the greater number of vegetable substances, and because the lead in this salt is partially thrown down by carbonic acid, if it happens to be present in the solution. Another precipitate is *nitrate of silver*, which throws down the bitter principle in light-yellow flocks, but this precipitant is also somewhat doubtful, for the same reason that renders acetate of lead so. The bitter principle of hops is likewise very soluble both in water and in alcohol.

Both the bitter principle and peculiar flavour of hops are communicated to wort. Much difference exists in the quantity employed, varying according to the taste of the persons who are to drink the ale. If the ale is to be strong, a greater quantity of hops can be employed without injury. Usually, the English brewers employ a much greater quantity of hops than the Scotch brewers.

Generally when the ale has considerable strength, the Edinburgh brewers are in the habit of adding one pound of hops for every bushel of malt employed. In fact, they sometimes, if they wish their ale to be very superior in flavour and quality, employ a greater quantity than even this. Thus it is seen that 100 pounds of hops are boiled in the strong ale wort extracted from 72 bushels of malt. If the ale is but weak, and consequently cheap, the usual allowance is one pound of hops to a bushel and a half of malt.

The peculiar flavour of the best ales is imparted by the skilful use of the hops; and both the quantity employed and the time of the boiling demand the mature consideration of the brewer. The best European hops for ale are the Kent growth, of a pale-green colour, glossy, and having an aromatic flavour. An arbitrary rule cannot be given for the quantity to be used in brewing ales of different strengths, as much depends on the views of the brewer with regard to the future disposal of the ale.

From 1 to 1½ pounds of hops per bushel is used in Edinburgh for the best strong ales, using a third more for the summer-keeping ale than for winter ale or ale brewed for immediate use. I believe the peculiar substances in hops connected with brewing have been already noticed. The ale is rendered very unpalatable if too much of the bitter principle is used. The Nottingham brewers, who are said to be the best in England, sometimes use as much as 15 pounds of hops to 9 bushels of malt. The porter-brewers of London use very little Kent hops of fine quality: they prefer the cheaper red hops of Sussex and other districts. Two pounds and a half per barrel of 36 gallons, made from three bushels of malt, is the quantity used by the English provincial brewers.

The wort should be transferred into the copper, and made to boil as soon as possible, for if it remains long in the underback it is apt to become acescent. The steam moreover, raised from it in the act of boiling, serves to screen it from the oxyginating or acidifying influence of the atmosphere. Until it begins to boil, the air should be excluded by some kind of a cover.

Sometimes the first wort is brewed by itself into strong ale, the second by itself into an intermediate quality; and the third into small-beer. But this practice is not much followed in this country.

After the wort has been boiled down to the requisite strength, which is commonly between the specific gravities 1·09 and 1·10, it is let out into the *coolers*. · The coolers are floors of wood, surrounded with a wooden ledge, and water-tight, placed in the most airy and exposed situation in the brewery. They are of such a size as to hold the whole of the wort at a depth not exceeding three or four inches, so that, in large breweries, they are of an enormous size. The intention is to cool down the wort as rapidly as possible to the temperature of the atmosphere; for, if it were allowed to remain long hot, it would run the risk of becoming sour, which would spoil the whole process. Very much of the superiority of some breweries over others depends upon the construction of the coolers, or rather upon their being as well adapted as possible for reducing the temperature of the wort speedily to that of the atmosphere. Free currents of air ought to pass over them, and great care should be taken to keep them perfectly clean, and occasionally let them be sweetened with limewater.

The wort is either pumped out of the boiler into the coolers, or let into them by simply opening a stopcock, according to the construction of the brewhouse. It soon spreads itself over all the surface of the coolers, and a very great evaporation is the consequence. This evapora-

tion should always be taken into consideration by the brewer, because it both materially adds to the strength of the ale
and diminishes its quantity. The amount of it depends
upon the temperature of the air compared with the dryness of the atmosphere, and upon the skill with which the
coolers have been constructed.

The hot wort reaches the cooler at a temperature of
from 200° to 208°. Here it should be cooled to the proper temperature for the fermenting tun, which may vary
from 54° to 64°, according to circumstances. The refrigeration is accomplished by the evaporation of a portion of the liquor : it is more rapid in proportion to the
extent of the surface, to the low temperature, and the
dryness of the atmosphere surrounding the cooler. The
renewal of a body of cool dry air by the agency of a fan,
may be employed with great advantage. The cooler itself
must be so placed that its surface shall be freely exposed
to the prevailing wind, and be as free as possible from the
eddy of surrounding buildings. It is thought by many,
that the agitation of the wort during its cooling is hurtful.
Were the roof made movable, so that the wort could be
readily exposed, in a clear night, to the aspect of the sky,
it would cool rapidly by evaporation, on the principles
explained by Dr. Wells in his "Essay on Dew."

If the cooling is effected by evaporation alone, the temperature falls very slowly, even in cold air, if it be loaded
with moisture. But when the air is dry, the evaporation
is vigorous, and the moisture exhaled does not remain in-
cumbent on the liquor, as in damp weather, but is diffused
widely in space. Hence we can understand how wort
cools so rapidly in the spring and autumn, when the air is

generally dry, and even more quickly than in winter, when the air is cooler, but loaded with moisture. In fact, the cooling process goes on better when the atmosphere is from 50° to 55° than when it falls to the freezing point; because in this case, if the air be still, the vapours generated remain on the surface of the liquor, and prevent further evaporation. In summer, the cooling can take place only during the night.

In consequence of the evaporation during this cooling process, the bulk of the worts is considerably reduced; thus, if the temperature at the beginning was 208°, and if it be at the end 64°, the quantity of water necessary to be evaporated to produce this refrigeration would be nearly ⅓ of the whole, putting radiation and conduction of heat out of the question. The effect of this will be a proportional concentration of the beer.

The period of refrigeration in a well-constructed cooler, amounts to six or seven hours in favourable weather, but to twelve or fifteen in other circumstances. The quality of the beer is much improved by shortening this period; because, in consequence of the great surface which the wort exposes to the air, it readily absorbs oxygen, and passes into the acetous fermentation with the production of various mouldy spots; an evil to which ill-hopped beer is particularly liable. Many schemes have been contrived to cool wort, by transmitting it through the convolutions of a pipe immersed in cold water. The best plan is to expose the hot wort for some hours freely to the atmosphere and the cooler, when the loss of heat is most rapid by evaporation and other means, and, when the temperature falls to 100°, or thereabout, to transmit the liquor through a zigzag pipe, laid almost horizontally in a trough of cold water.

When the wort is reposing in the cooler, it lets fall a slight sediment, which consists partly of fine flocks of co-agulated albumen combined with tannin, and partly of starch, which had been dissolved at the high temperature, and separates at the lower. The wort should be perfectly limpid, for a muddy liquor never produces transparent beer. Beer·of this kind contains, besides mucilaginous matter, sugar, and gum, usually some starch, which even remains after the fermentation, and hinders its clarifying, and gives it a tendency to sour. The wort contains more starch the hotter it has been mashed, the less hops that have been added, and the shorter time it has been boiled. The presence of starch in the wort may be made manifest by adding a little *tincture of iodine* to it, when it will become immediately blue. It is thus seen that the tranquil cool-ing of wort in a proper vessel has an advantage over cool-ing it rapidly by a refrigeratory apparatus. When the wort is sufficiently cool, it is let down into the fermenting-tun. In this transfer the cooling might be carried several degrees lower, were the wort made to pass down through a tube enclosed in another tube, along which a stream of cold water is flowing in the opposite direction. These fermenting tuns are commonly called *gyle-tuns,* or working tuns, and are either square or circular, the latter being preferable on many accounts. They are of very great size in the London breweries, but in private brew-houses they often do not exceed the size of a wine-hogshead, or even of a beer-barrel. The fermentation is perhaps conducted with the greatest economy in large vessels; but good ale may be made in comparatively small quantities. To what extent this is the case with porter, it is more difficult to say. Fine porter has scarcely ever been made exept by those who manufacture it upon a large scale.

The fermenting tuns are not to be entirely filled by the wort, because a considerable increase in bulk takes place during the fermentation, in consequence of which the liquor would run over, unless allowance were made for it.

Fermentation of ale or beer is never carried to any great length. The object of the brewer is to retain the flavour and good qualities of the ale or beer, not to develop the greatest quantity of spirits, which can hardly be done without allowing the wort to run into acidity. The activity of the fermentation depends upon the quantity of yeast added. Accordingly, brewers mix yeast with their worts only in very sparing quantities, while the distiller adds it in large quantities and very often.

Yeast is a frothy substance, of a brownish-gray colour and bitter taste, which is formed on the surface of ale or wine while fermenting. If it be put into sacks, the moisture gradually drops out, and the yeast remains behind in a solid form. It has very much of the flavour and taste of cheese when in this state; but its colour is still darker. This dried yeast promotes or excites fermentation, but it does not answer quite so well as fresh yeast. From the resemblance which dried yeast has to cheese, we would be naturally inclined to infer that it is a species or variety of gluten. But if we attempt to induce fermentation in wort by adding the gluten of wheat, we will be unsuccessful.

After yeast is kept for some time in a cylindrical glass

vessel, a white substance, not unlike curd, separates and swims on the surface. If this substance be removed, the yeast loses the property of exciting fermentation. This white substance possesses many of the properties of gluten, or vegetable *fibrin*, though it differs from it in others. Its colour is much whiter; it has not the same elasticity, and its particles do not adhere with the same force. In short; it agrees much more nearly, in its properties, with curd of milk, than with gluten of wheat.

Dobereiner found that when yeast is steeped in alcohol it loses the property of acting as a ferment. This may be owing to the alcohol dissolving and carrying off the true fermenting principle.

It does not seem unlikely that the portion of yeast which really acts as a ferment is a quantity of saccharine matter which it contains, that has begun to undergo the decomposition produced by fermentation, but has not yet completed the change. It is well known that nothing more seems to be necessary than to begin the fermentative process in wort; the process then goes on of itself. The curious have been anxious to know whether a high temperature (96° or 100°) might not be substituted in distilleries for the great quantities of yeast at present employed.

There is but a very small quantity of yeast mixed with the wort in the fermenting process, about a gallon of yeast for every three barrels of wort. Very soon after the yeast has been mixed with the wort, an intestine motion begins to appear in the liquid; air-bubbles separate from it, and a froth collects slowly upon the surface. This froth is of a yellowish-gray colour. It has the appearance of cream at first; but in a few days it collects in considerable quantities, especially if the weather be warm. At the same time, the temperature of the wort increases, and a very considerable quantity of carbonic acid gas is given out by it. The increase of temperature which takes place during the fermenting of ale may be stated, on an average, to amount to 12° or 15°. Sometimes it amounts to 20°, and sometimes does not exceed 5°. But, in such cases, there is generally some fault in the skill of the brewer. In slow fermentations, the degree of heat at which the wort is pitched is from 50° to 55°. When the gyle comes to a head, the yeast that forms on the surface is beat in, and this process continues from nine to twelve days. As soon as the ale is judged to be sufficiently attenuated, it is run from beneath the yeast into barrels.

The wort, in quick fermentation, is mixed with yeast at from 60° to 65° of heat. The yeast is permitted to form on the head of the gyle, and gives out carbonic acid until it begins to turn viscid, and tends to sink down through the wort. This takes place in from thirty-six to forty-eight hours. It is then mixed well in the gyle, and run into barrels, where it ferments and works over for forty-eight hours, requiring to be regularly filled up, at first every two hours for the first twelve hours, then every four hours, until it gradually comes to yeast, and ceases working out of the barrels.

There are two distinct methods of fermentation. The quick fermentation is that by which porter is made, and ale generally, throughout England. The slow fermentation is the practice in brewing ale in Edinburgh and in Scotland. In both methods, the decomposition of the glucosin and the formation of alcohol go on until the ale is finished,

4*

and either run into barrels, as in Edinburgh, or until the fermentation ceases in the barrels, as in England. This latter is the proper term for *cleansing*. In general, English brewers know nothing about attenuation. Their practice of fermenting at a high heat forces on the gyle so rapidly, that as soon as ready, which is within forty-eight hours, they must run the wort into barrels, to check the heat and bring it to cleanse into yeast, which does not begin to form until twelve or fourteen hours after the wort is first run into barrels. The repeated filling of the barrels as the wort flows out, being equivalent to the beating in of the head of the yeast for some days by the Scotch brewers.

Many of the Edinburgh brewers attenuate their high-priced bottled ales as low as they can carry it with safety, for the purpose of making it keep. The weak draught ale is not attenuated so much. This gives it a fulness to the taste.

Cleansing is a name given to the last of the processes of brewing. When the violence of the fermentation is over, the head of yeast which covers the top of the fermenting-tun diminishes in height by the gradual escape of carbonic acid gas, which heaved it into bubbles. If the wort were allowed to remain in the gyle-tun after this has happened, the yeast would again mix with it; and the consequence would be a disagreeable bitter taste, known among brewers by the name of "*yeast bitter.*" The fermentation would likewise continue, though in a languid manner, and the ale would soon run into acidity. These accidents are prevented by drawing off the ale into casks, and this is called "*cleansing.*"

The casks are filled quite full, and left with their bungs open. The temperature of the ale is lowered, by drawing off from the gyle-tun, which lowers its temperature and checks its fermentation. For this reason, the cleansing is sometimes practised in summer, when the elevation of temperature in the wort is at its height.

It has been repeatedly observed that a curious circumstance takes place during the cleansing, not very easily accounted for. If we take the temperature of the ale at the upper surface of the gyle-tun, and then observe the temperature of the ale when it flows from the stopcock at the bottom of the tun, we shall generally find it one or two degrees hotter in this latter place than at the former. Very naturally, we ought to expect the highest temperature at the top of the gyle-tun.

The ale still continues to ferment after it is put into the small casks; but as these casks are always kept full, the yeast, as it comes to the surface, flows out at the bung, and thus separates altogether from the beer. It is this separation that has induced brewers to distinguish it by the name of *cleansing*. In these casks, then, the yeast divides itself into two portions. The greater part rises up with the carbonic acid evolved, and flows out at the bunghole; while another portion subsides to the bottom, and constitutes what is called the dregs of the beer. It is essential to the cleansing that the casks should be always full—otherwise the yeast will not run off, and the beer will not become transparent. This object is accomplished in small breweries by a man constantly going round and filling up the casks as they *work down*. In the London breweries there is an ingenious mechanical contrivance which answers the purpose perfectly.

The beer will usually be clear as soon as the fermentation has subsided. It is bunged up in the casks, and preserved for sale; or in London, where the quantity is too great for this, the beer is removed into large stone vats, capable of holding several thousand barrels, from which it is gradually distributed to the consumers.

The beer is sent to the public-houses in London, commonly, as soon as fermentation is over, and before it has had time to become fine. It is usual to send along with it a quantity of *finings*, as it is called; that is, a solution of isinglass in weak sour beer, made from a fourth mash of the same malt. A certain quantity of this is put in every cask by the publican. It forms a kind of web at the surface of the liquid; and, gradually sinking to the bottom, carries with it all the flocculent matter, and leaves the beer transparent.

ALE AND BEER.

THESE two words, in Great Britain and this country, are applied to two liquors obtained by fermentation from the malt of barley; but they differ from each other in several particulars. Ale is light-coloured, brisk, and sweetish, or at least free from bitter; while beer is dark-coloured, bitter, and much less brisk. What is called *porter*, in England, is a species of beer; and the term porter at present signifies what was formerly called *strong beer*. The original difference between these two liquids was owing to the malt from which they were prepared. Ale malt was dried at a very low heat, and consequently was of a pale colour; while beer or porter malt was dried at a higher temperature, and had of consequence acquired a brown colour. This incipient charring had developed a peculiar and agreeable bitter taste, which was communicated to the beer along with the dark colour. This bitter taste renders beer more agreeable to the palate, and less injurious to the constitution, than ale. It was consequently manufactured in greater quantities, and soon became the common drink of the lower ranks, wherever it was introduced.

When malt became high-priced, in consequence of the heavy taxes laid upon it, and the great increase in the price of barley which took place during the war of the French revolution, the brewers found out that a greater quantity of wort of a given strength could be prepared from pale malt than from brown malt. The consequence was, that pale malt was substituted for brown malt in the brewing of porter and beer.

It is not to be understood that the whole malt employed was pale, but the greater proportion of it. The wort, of course, was much paler than before, and it wanted that agreeable bitter flavour which characterized porter and made it so much relished by most palates. Several endeavours were made by brewers by artificial additions to remedy these defects. They prepared an artificial colouring matter, by heating a solution of coarse sugar in an iron boiler till it became black, and was reduced to the consistency of treacle. The smoke issuing from it was then set on fire, and the whole was allowed to burn for about ten minutes, when the flame was extinguished by putting a lid on the vessel. A certain quantity of water was mixed with this substance before it was cold. The porter is coloured by adding about two pounds of this colouring matter for every barrel of wort, while in the copper. Some brewers make their colouring matter with infusion of malt instead of sugar; and, in 1809, M. de Roche took out a patent for preparing the colouring matter from the husks of malt, by burning them like coffee, and then infusing them in water. It is believed that all these colouring matters are of the same nature : of course, the brewer ought to employ that one of them which is cheapest.

To supply the place of the agreeable bitter which was communicated to porter by the use of brown malt, various substitutes were tried. Quassia, cocculus indicus, and, we believe, even opium, were employed in succession; but none of them were found to answer the purpose sufficiently.

The usual limits of the wort of strong ale may be stated at from 60 to 120 pounds per barrel, or from the specific gravity 1·064 to 1·11275 at the temperature of 60°. The highest-priced ales also are not always the strongest, because the price depends in a great measure on the reputation of the brewer. The fermentation of ale is not carried far; and the consequence is that a considerable portion of the saccharine matter still remains in the liquid, apparently unaltered. Traces of starch may be still detected in strong ale, even after it has been kept some time in bottles, by means of an infusion of nut-galls.

No arbitrary rule, however, can be laid down for the attenuation of the wort during the process of fermentation, as that must depend on the views of the brewer. Within the last five years, the Edinburgh brewers have not carried the attenuation of their ales, especially of that made to be used on draught, so far down as formerly. But in all cases, regard must be had to the time the ale is intended to be kept, and the season of the year.

Strong ale is much stronger than most specimens of porter. The average specific gravity of porter wort, according to Shannon, as deduced by the *saccharometer*, is 1·0645, which indicates 60 pounds per barrel of saccharine extract. Thus we see the reason why it is so much less glutinous and adhesive than strong ale. Porter undergoes much less fermentation than ale, according to the statement of many who have written on this subject; but we have not very accurate information on the subject. According to the experiments of Mr. Brande, brown stout, which is the strongest porter made in London, contains 6·8 per cent. by measure, of alcohol of the specific gravity 0·825.

There are three kinds of malt used by the brewers in London, to wit, pale malt, amber malt, and brown malt. Each of these is mashed separately, and the worts from each are afterward mixed together in the same fermenting vessel. In some breweries there are three separate mash-tuns. The custom, in other breweries, is to mash one kind of malt the first day, another kind the second day, and a third kind the third day. The first day's wort is put into the fermenting-vessel, and mixed with yeast; and the other two worts are added to it successively as

they are formed. It will thus be perceived that it is very
difficult to determine with accuracy the strength of the
worts in most breweries. It could only be done by
knowing the quantity of wort from each mash, and its specific gravity when let into the fermenting-vessel. Only
two or three days are occupied in the fermentation of
porter, so the process is carried through rapidly.

Various proportions of pale and brown malt are used
in different breweries. One of the best breweries in London is said to use nearly two parts of pale malt to one
part of brown.

PLAN, MACHINERY, AND UTENSILS OF A GREAT BREWERY.

FIGS. 7 and 8 represent the arrangement of the utensils
and machinery in a porter brewery on the largest scale;
in which however it must be observed that the elevation,
fig. 7, is in a great degree imaginary as to the plane upon
which it is taken; but the different vessels are arranged
so as to explain their uses most readily, and at the same
time to preserve, as nearly as possible, the relative posi-
tions which are usually assigned to each in works of this
nature.

The malt for the supply of the brewery is stored in vast
granaries or malt-lofts, usually situated in the upper part
of the buildings. Of these only one has been represented
at A, fig. 7; the others, which are supposed to be on each
side of it, cannot be seen in this view. Immediately be-

neath the granary A, on the ground-floor, is the mill; in
the upper story above it, are two pairs of rollers, figs. 5,

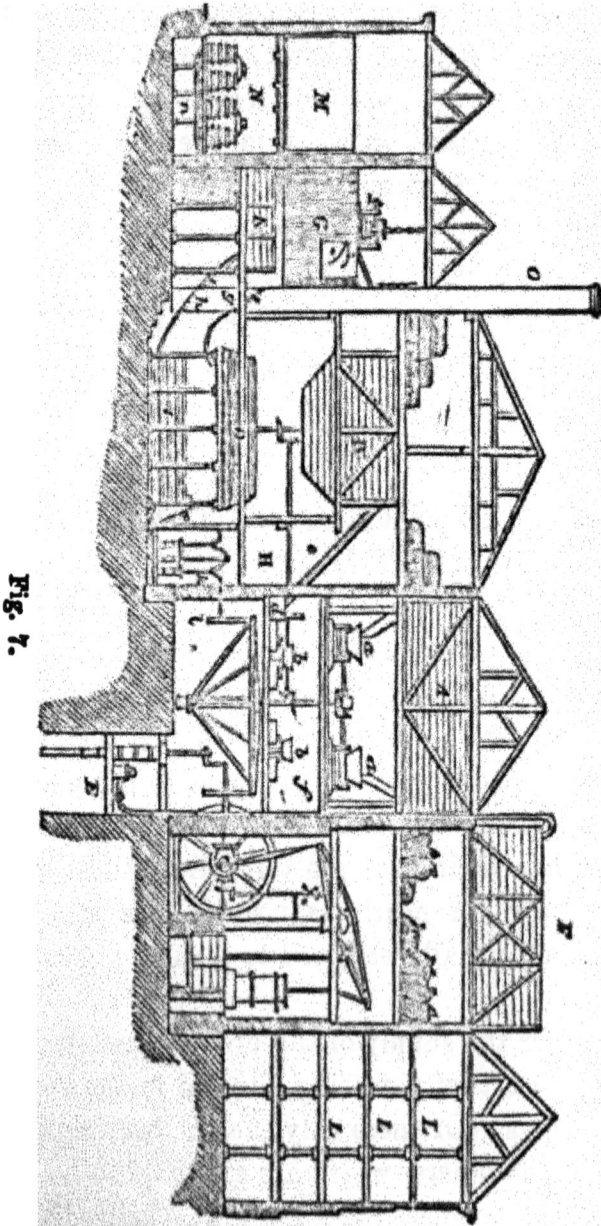

Fig. 7.

6, and 7, under *a*, *a*, for bruising or crushing the grains of the malt. In the floor, beneath the rollers, are the millstones *b*, *b*, where the malt is sometimes ground, instead of being merely bruised by passing between the rollers, under *a*, *a*.

When the malt is prepared, it is conveyed by a trough into a chest *d*, to the right of *b*, from which it can be elevated by the action of a spiral screw, fig. 8, enclosed in the sloping tube *e*, into the large chest or bin B, for holding ground malt, situated immediately over the mash-tun D. The malt is reserved in this bin till wanted, and it is then let down into the mashing-tun, where the extract is obtained by hot water supplied from the copper G, seen to the right of B.

The water for the service of the brewery is obtained from the well E, seen beneath the mill to the left, by a lifting-pump worked by a steam-engine; and the forcing-pipe *f* of this pump conveys the water up to the large reservoir or waterback F, placed at the top of the engine-house. From this cistern, iron pipes are laid to the copper G, (on the right hand side of the figure,) as also to every part of the establishment where cold water can be wanted for cleansing and washing the vessels. The copper G can be filled with cold water by merely turning a cock; and the water, when boiled therein, is conveyed by the pipe *g* into the bottom of the mash-tun D. It is introduced beneath a false bottom, upon which the malt lies, and, rising up through the holes in the false bottom, it extracts the saccharine matter from the malt; a greater or less time being allowed for the infusion, according to circumstances. As soon as the water is drawn off from the copper, fresh water must be let into it, in order to be ready for boiling the second mashing; because the copper must not be left empty for a moment—otherwise the intense heat of the fire would destroy its bottom.

For the convenience of thus letting down at once as much liquor as will fill the lower part of the copper, a pan or second boiler is placed over the top of the copper, as seen in fig. 7, and the steam rising from the copper communicates a considerable degree of heat to the contents of the pan, without any expense of fuel. This will be more minutely described further on. (See fig. 11.)

While the process of mashing is going on, the malt is agitated in the mash-tun so as to expose every part to the action of the water. This is done by a mechanism contained within the mash-tun, which is put in motion by a horizontal shaft above it, H, leading from the mill. The mash-machine is shown separately in fig. 10. When the operation of mashing is completed, the wort or extract is drained down from the malt into the vessel I, called the underback, immediately below the mash-tun, of like dimensions, and situated always on a lower level, for which reason it has received this name.

The wort does not remain here longer than is necessary to drain off the whole of it from the tun above. Then it is pumped up by the three-barrelled pump k, into the pan above the top of the copper, by a pipe which cannot be seen in this section. The wort remains in the pan until the water for the succeeding mash is discharged from the copper. But this delay is no loss of time, because the heat of the copper and the steam arising from it prepare the wort, which had become cooler, for boiling.

Immediately after the copper is empty, the first wort is let down from the pan into the copper, and the second wort is pumped up from the underback into the upper pan. The proper proportion of hops is thrown into the copper through the near hole, and then the door is shut down and screwed fast, to keep in the steam, and cause it to rise up through pipes into the pan. Thus it is forced to blow up through the wort in the pan, and communicates so much heat to it, or water, called *liquor* by the brewers, that either is brought near to the boiling-point. The different worts succeed each other through all the different vessels with the greatest regularity, so that there is no loss of time, but every part of the apparatus is constantly employed. When the boiling has continued a sufficient period to coagulate the grosser part of the extract, and to evaporate part of the water, the contents in the copper are run off through a large cock into the *jackback* K, below G, which is a vessel of sufficient dimensions to contain it, and provided with a bottom of cast-iron plates, perforated with small holes, through which the wort drains and leaves the hops.

The hot wort is drawn off from the jackback through the pipe *h*, by the three-barrelled pump, which throws it up to the coolers L, L, L; this pump being made with different pipes and cocks of communication, to serve all the purposes of the brewery except that of raising the cold water from the well.

The coolers L, L, L are very shallow vessels, built over one another in several stages; and that part of the building in which they are contained is built with lattice-work or shutter-flaps, on all sides, to admit free currents of air.

After the wort is sufficiently cool to put to the first fermentation, it is carried through pipes from all the different
coolers to the large fermenting-vessel or gyle-tun M, which,
with another similar vessel behind it, is of sufficient capacity to contain all the beer of one day's brewing.

Whenever the first fermentation is concluded, the beer
is drawn off from the great fermenting-vessel M, into the
small fermenting-casks, or cleansing-vessels N, of which
there are a great number in the brewery. They are placed
four together, and to each four a common spout is provided to carry off the yeast and conduct it into the
troughs u, placed beneath. The beer remains in these
cleansing-vessels till the fermentation is completed; and
it is then put into the store-vats, which are casks or tuns
of immense size, where it is kept till wanted, and is finally drawn off into barrels and sent away from the brewery.
The store-vats are not represented in the figure. They are
of a conical shape and of different dimensions, from fifteen
to twenty feet diameter, and usually from fifteen to twenty feet in depth. The steam-engine which puts all the
machinery in motion is exhibited in its place, on the left
side of the figure. On the axis of the large fly-wheel is
a bevelled spur-wheel, which turns another similar wheel
upon the end of a horizontal shaft, which extends from
the engine-house to the great horse-wheel, set in motion
by means of a spur-wheel. The horse-wheel drives all
the pinions for the millstones b, b, and also the horizontal
axis which works the three-barrelled pump k. The rollers a, a are turned by a bevel wheel upon the upper end
of the axis of the horse-wheel, which is prolonged for
that purpose; and the horizontal shalft H, for the mash-

5*

ing-engine is driven by a pair of bevel wheels. There is likewise a sack-tackle, which is not represented. It is a machine for drawing up the sacks of malt from the court-yard to the highest part of the building, whence the sacks are wheeled on a truck to the malt-loft A, and the con-tents of the sacks are discharged.

The horse-wheel is intended to be driven by horses occasionally, if the steam-engine should fail; though such is not often the case at the present day, as the steam-en-gine is brought to great perfection.

Fig. 8 is a representation of the *fermenting-house* at the brewery of Messrs. Whitbread and Company, Chiswell street, London, which is said to be one of the most complete in its arrangement in the world: it was erected after the plans of Mr. Richardson, who conducts the brewing at those works. The whole of fig. 8 is to be considered as devoted to the same object as the large vessel M, and the cask N, fig. 7. In fig. 8, *r r* is the pipe which leads from the dif-ferent coolers to convey the wort to the great fermenting vessels or squares M, of which there are two, one behind the other; *f f* represents a part of the great pipe which conveys all the water from the well E, fig. 7, up to the water-cistern F. This pipe is conducted intentionally up the wall of the fermenting-house, fig. 8, and has a cock in it, near *r*, to stop the passage. Immediately beneath this passage a branch-pipe *p* proceeds and enters a large pipe *x x*, which has the former pipe *r* withinside of it. From the end of the pipe *x*, nearest to the squares M, an-other branch *n n* proceeds, and returns to the original pipes, with a cock to regulate it. The object of this ar-rangement is to make all or any part of the cold water

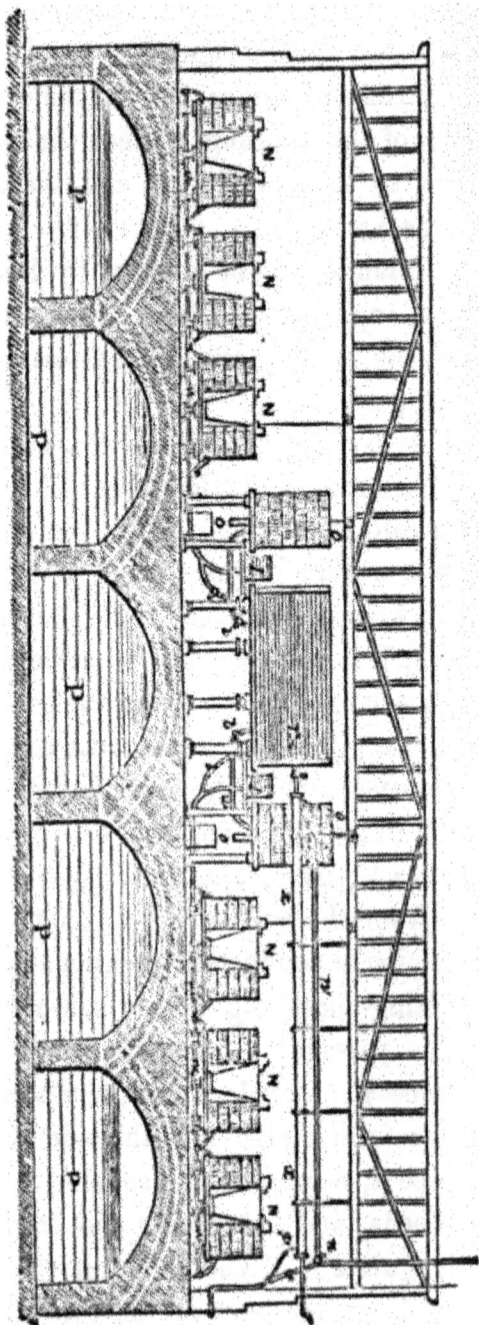

Fig. 8.

flow through the pipe *x x*, which surrounds the pipe *r*, formed only of thin copper, and thus cool the wort passing through the pipe *r* until it is found by the thermometer to have the exact temperature which is desirable before it is put to ferment in the great square M. By means of the cocks at *r* and *p*, the quantity of cold water passing over the surface of the pipe *r* can be regulated at pleasure, whereby the heat of the wort, when it enters into the square, may be adjusted within half a degree.

When the first fermentation in the squares M, M is finished, the beer is drawn off from them by pipes marked *v*, and conducted by their branches w, w, w, to the different rows of fermenting-tuns, marked N, N, which occupy the greater part of the building. In the hollow between every two rows are placed large troughs to contain the yeast which they throw off. The figure shows that the small tuns are all placed on a lower level than the bottom of the greatest vessel M, so that the beer will flow into them, and, by hydrostatic equilibrium, will fill them to the same level. When they are filled, the communication-cock is shut; but as the working off of the yeast diminishes the quantity of beer in each vessel, it is necessary to replenish them from time to time. For this purpose the two large vats *o, o* are filled from the great squares M, M, before any beer is drawn off into the small cask N, and this quantity of beer is reserved at the higher level for filling up. The two vessels *o, o* are, in reality, situated between the two squares M, M; but they have necessarily been placed thus in this section, in order that they may be seen.

Near each filling-up tun *o* is a small cistern I, communicating with the tun *o* by a pipe, which is closed by a float-valve. The small cisterns *t* are always in communication with the pipes which lead to the small fermenting-vessel N; and therefore the surface of the beer in all the tuns and in the cisterns will always be at the same level; and as this level subsides by the working off of the yeast from the tuns, the float sinks and opens the valve, so as to admit a sufficiency of beer from the filling-up tuns *o*, to restore the surface of the beer in all the tuns, and also in the cistern *t*, to the original level.

In order to carry off the yeast which is produced by the fermentation of the beer in the tuns *o, o*, a conical iron dish or funnel is made to float upon the surface of the beer which they contain; and from the centre of this funnel a pipe *o* descends, and passes through the bottom of the tun, being packed with a collar of leather, so as to be water-tight; at the same time that it is at liberty to slide down, as the surface of the beer descends into the tun. The yeast flows over the edge of this funnel-shaped dish, and is conveyed down the pipe to a trough beneath.

Beneath the fermenting-house are large arched vaults P, built with stone, and lined with stucco. The beer is let down into these in casks, as soon as it is sufficiently fermented, and is there kept until called for. These vaults are used at Mr. Whitbread's brewery, instead of the great stone vats which have been before spoken of, and are in some respects preferable, because they preserve a greater equality of temperature, being beneath the surface of the earth.

The malt-rollers, or machines for bruising the grains of the malt, (figs. 5, 6,) have been already described.

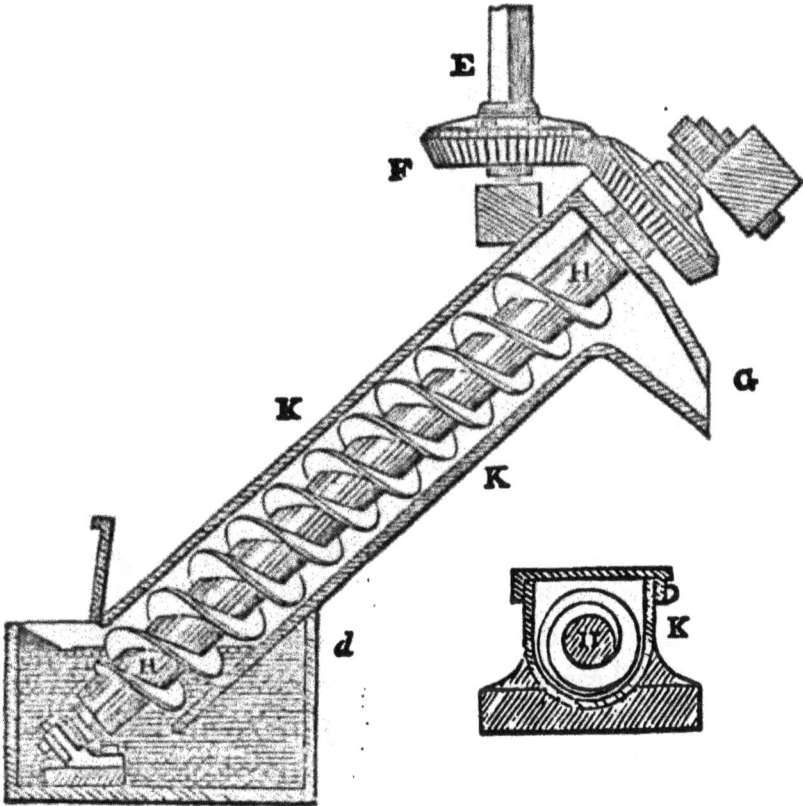

Fig. 9.

The malt is shot down from A, fig. 7, the malt-loft, into the hopper; and from this it is let out gradually through a sluice or sliding shuttle *a*, fig. 7, and falls between the rollers.

Fig. 9 is the screw by which the ground or bruised malt is raised up, or conveyed from one part of the brewery to another. K is an inclined box or trough, in the centre of which the axis of the screw H is placed; the spiral iron plate or worm which is fixed projecting from the axis, and which forms the screw, is made very nearly to fill the inside of the box. By this means, when the screw is turned round by the wheels E, F, or by any other means, it raises up the malt from the box d, and delivers it at the spout G.

This screw is equally applicable for conveying the malt horizontally in a trough K as slantingly; and similar machines are employed in various parts of breweries, for conveying the malt wherever the situation of the works requires.

Fig. 10 is the mashing-machine. $a\ a$ is the tun, made of wood staves hooped together. In the centre of it rises a perpendicular shaft b, which is turned slowly round by means of the bevelled wheels t, u at the top. c, c are two arms, projecting from the axis, and supporting the short vertical axis d of the spur-wheel x, which is turned by the spur-wheel w; so that when the central axis b is made to revolve, it will carry the thick short axle d round the tun in a circle. That axle d is furnished with a number of arms c, c, which have blades placed obliquely to the plane of their motion. When the axis is turned round, these arms agitate the malt in the tun, and give it a constant tendency to rise upward from the bottom.

The motion of the axle d is produced by a wheel x, on the upper end of it, which is turned by a wheel w, fastened on the middle of the tube b, which turns freely round upon its central axis. Upon a higher point of the same tube b is a bevel wheel o, receiving motion from a bevel wheel q, fixed upon the end of the horizontal axis $n\ n$, which gives motion to the whole machine. This same axis has a pinion p upon it, which gives motion to the wheel r, fixed near the middle of a horizontal axle,

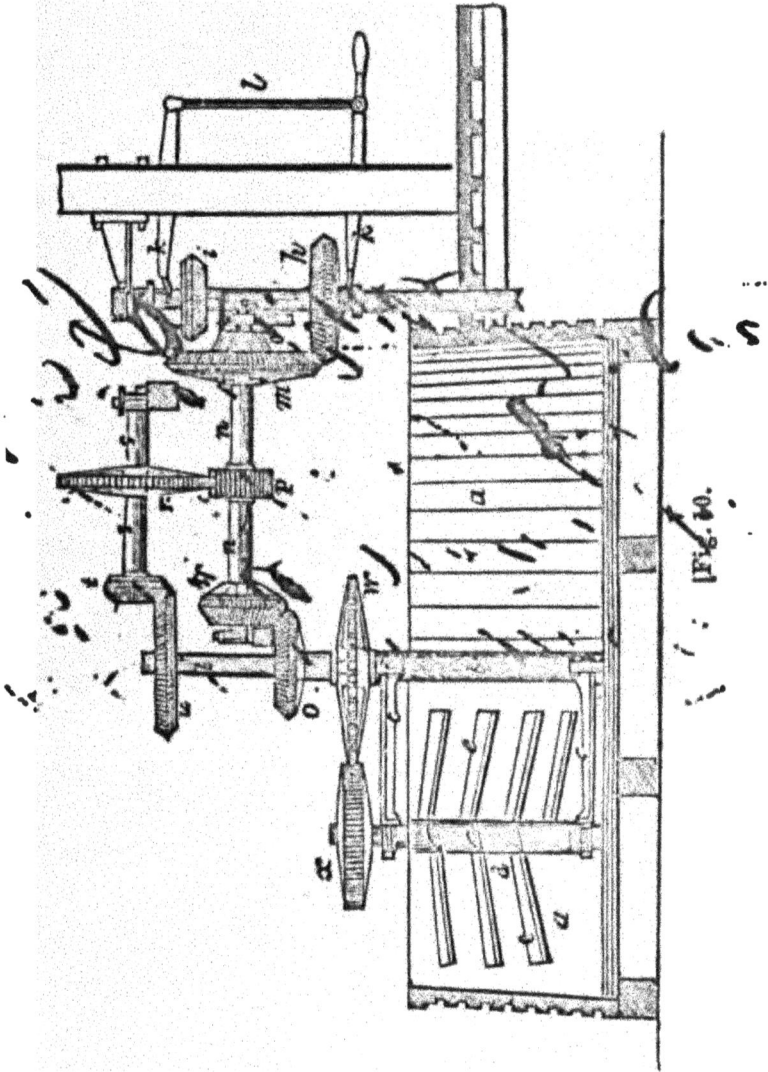

Fig. 10.

which, at its left-hand end, has a bevel pinion *t*, working the wheel *u*, before mentioned. By these means, the rotation of the central axis *b* will be very slow, compared with the motion of the axle *d*; for the latter will make seventeen or eighteen revolutions on its own axis in the same space of time that it will be carried once round the tun by the motion of the shaft b. At the beginning of the operation of mashing, the machine is made to turn with a slow motion; but, after having wetted all the malt by one revolution, it is driven quicker. For this purpose, the ascending shaft fg, which gives motion to the machine, has two bevel wheels h, i, fixed upon a tube fg, which is fitted upon a central shaft.

These wheels actuate the wheels m and o, upon the end of the horizontal shaft n n; but the distance between the two wheels h and i is such that they cannot be engaged both at once with the wheels m and o; but the tube fg, to which they are fixed, is capable of sliding up and down on its central axis sufficiently to bring either wheel h or i into gear with its corresponding wheel o or m, upon the horizontal shaft; and as the diameters of n o and i m are of very different proportions, the velocity of the motion of the machine can be varied at pleasure, by using one or the other. k and k are two levers, which are forked at their extremities, and embrace collars at the ends of the tube fg. These levers being united by a rod l, the handle k gives the means of moving the tube fg, and its wheels h, i, up or down, to throw either the one or the other wheel into gear.

The object of boiling the wort is not merely evaporation and concentration, but extraction, coagulation, and, finally, combination with the hops; purposes which are better accomplished in a deep confined copper, by a moderate heat, than in an open shallow pan with a quick fire. The copper, being incased above in brickwork, retains its digesting temperature much longer than the pan could do. The waste steam of the close kettle, moreover, can be economically employed in communicating heat to water or weak worts; whereas the exhalations from an open pan would prove a nuisance, and would need to be carried off by a hood. The boiling has a fourfold effect: 1. It concentrates the wort; 2. During the earlier stages of heating, it converts the starch into sugar, dextrine, and gum, by means of the diastase; 3. It extracts the substance of the hops diffused through the wort; 4. It coagulates the albuminous matter present in the grain, or precipitates it by means of the tannin of the hops.

The degree of evaporation is regulated by the nature of the wort and the quality of the beer. Strong ale and stout for keeping, require more boiling than ordinary porter or table-beer brewed for immediate use. The proportion of the water carried off by evaporation is usually from one-seventh to one-sixth of the volume. The hops are introduced during the progress of the ebullition. They serve to give the beer not only a bitter aromatic taste, but also a keeping quality, or they counteract its natural tendency to become sour; an effect partly due to the precipitation of the albumen and starch, by their resinous and tanning constituents, and partly to the antifermentable properties of their lupuline, bitter principle, ethereous oil, and resin. In these respects, there is none of the bitter plants which can be substituted for hops with advantage. For strong beer, powerful fresh hops should be selected; for weaker beer, an older and weaker article will suffice.

The hops are either boiled with the whole body of the wort, or extracted with a portion of it, and this concentrated extract added to the rest. The stronger the hops are, the longer time they require for extraction of their virtues; for strong hops, an hour and a half or two hours boiling may be proper; for a weaker sort, half an hour or an hour may be sufficient; but it is never advisable to push this process too far, lest a disagreeable bitterness, without aroma, be imparted to the beer. In most breweries, it is the practice to boil the hops with a part of the wort, and to filter the decoction through a drainer, called the *jack hopback*. The proportion of hops to malt is very various; but, in general, from a pound and a quarter to a pound and a half of the former are taken for one hundred pounds of the latter in making good table-beer. For porter and strong ale, two pounds of hops are used, or even more: for instance, one pound of hops to a bushel of malt, if the beer be destined for the consumption of India or other warm climates.

During the boiling of the two ingredients, much coagulated albuminous matter, in various states of combination, makes its appearance in the liquid, constituting what is called the *breaking* or *curdling of the wort*, when numerous minute flocks are seen floating in it. The resinous, bitter, and oily-ethereous principles of the hops combine with the sugar and gum, or dextrine of the wort; but for this effect they require time and heat; showing that the boiling is not a process of mere evaporation, but one of chemical reaction. A yellowish-green pellicle of hop-oil and resin appears upon the surface of the boiling wort, in a somewhat frothy form: when this disappears, the boiling is presumed to be completed, and the beer is strained off into the cooler. The residuary hops may be pressed and used for an inferior quality of beer; or they may be

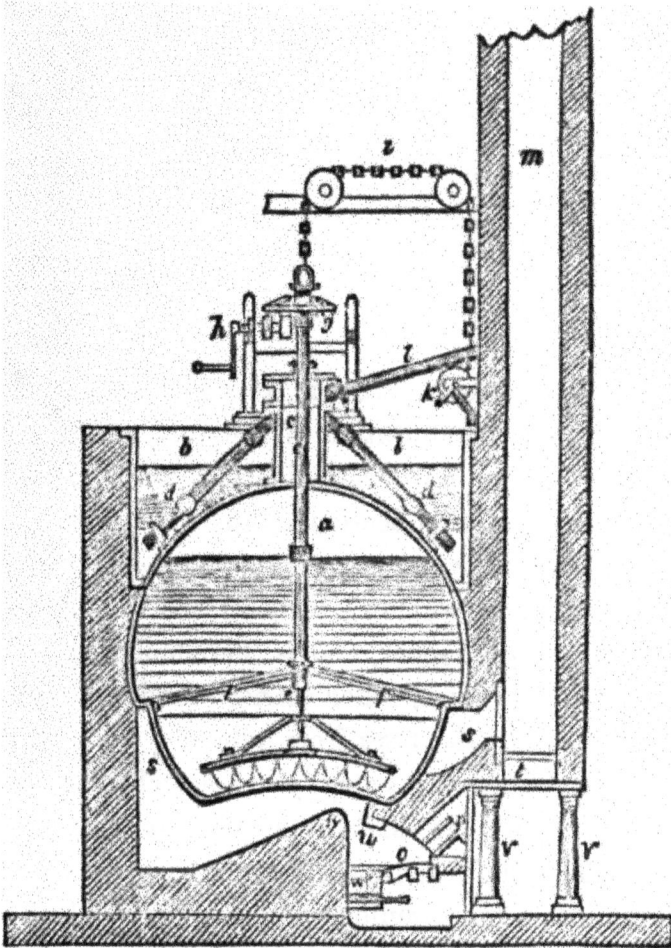

Fig. 11.

boiled with fresh wort, and be added to the next brewing charge.

Figs. 11, 12, represent the copper of a London brewery. Fig. 11 is a vertical section; fig. 12 a ground-plan of the fire-grate and flue, upon a smaller scale: *a* is the close copper kettle, having its bottom convex within; *b* is the open pan placed upon its top. From the upper part of

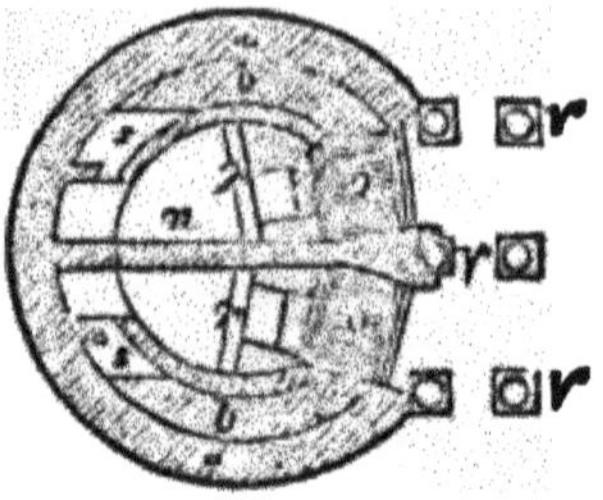

Fig. 12.

the copper, a wide tube c ascends, to carry off the steam generated during the ebullition of the wort, which is conducted through four downward-slanting tubes d, d, (two only are visible in this section,) into the liquor of the pan b, in order to warm its contents. A vertical iron shaft or spindle e passes down through the tube c, nearly to the bottom of the copper, and is there mounted with an iron arm, called a *rouser*, which carries round a chain hung in loops, to prevent the hops from adhering to the bottom of the boiler. Three bent stays f are stretched across the interior, to support the shaft by a collet at their middle junction. The shaft carries at its upper end a bevel wheel g, working into a bevel pinion upon the axis h, which may be turned either by power or by hand. The *rouser* shaft may be lifted by means of the chain i, which, going over two pulleys, has its end passed round the wheel and axle k, and is turned by a winch : l is a tube for conveying the waste steam into the chimney m.

The heat is applied as follows :—For heating the colossal coppers of the London breweries, two separate fires are required, which are separated by a narrow wall of brickwork, n, figs. 11, 12. The dotted circle $a' a'$ indicates the largest circumference of the copper, and $b'b'$ its bottom; o, o are the grates upon which the coals are thrown, not through folding doors, (as of old,) but through a short slanting iron hopper, shown at p, fig. 11, built in the wall, and kept constantly filled with the fuel, in order to exclude the air. Thus the lower stratum of coals gets ignited before it reaches the grate. Above the hopper p a narrow channel is provided for the admission of atmospherical air, in such quantity merely as may be requisite to complete the combustion of the smoke of the coals. Behind each grate there is a fire-bridge r, which reflects the flame upward, and causes it to play upon the bottom of the copper. The burnt air then passes round the copper in a semicircular flue $s\,s$, from which it flows off into the chimney m, on whose under end a sliding damper-plate t is placed for tempering the draught. When cold air is admitted at this orifice, the combustion of the fuel is immediately checked. There is, besides, another slide-plate at the entrance of the slanting-flue into the vertical chimney, for regulating the play of the flame under and around the copper. If the plate t be opened, and the other plate shut, the power of the fire is suspended, as it ought to be, at the time of emptying the copper. Immediately over the grate is a brick arch u, to protect the front edge of the copper from the first impulsion of the flame. The chimney is supported upon iron pillars, v, v; w is a cavity closed with a slide-plate, through which the ashes may be taken out from behind, by means of a long iron hook.

Fig. 13 represents one of the sluicecocks which are used to make the communications of the pipes with the pumps, or other parts of the brewery. B B represents the

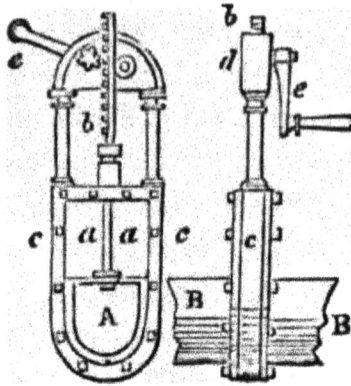

Fig. 13.

pipe in which the cock is placed. The two parts of this
pipe are screwed to the side of a box c c, in which a slider
A rises and falls, and intercepts, at pleasure, the passage
of the pipe. The slider is moved by the rod a. This
passes through a stuffing-box, in the top of the box which
contains the slider, and has the rack b fastened to it. The
rack is moved by a pinion fixed upon the axis of a handle
e, and the rack and pinion are contained in a frame d,
which is supported by two pillars. The frame contains a
small roller behind the rack, which bears it up toward the
pinion, and keeps its teeth up to the teeth of the pinion.
The slider A is made to fit accurately against the internal
surface of the box c, and to bear against this surface by
the pressure of a spring, so as to make a perfectly close
fitting.

Fig. 14 is a small cock to be placed in the side of the
great store-vats, for the purpose of drawing off a small
quantity of beer, to taste and try its quality. A is a part
of the stave or thickness of the great store-vat; into this

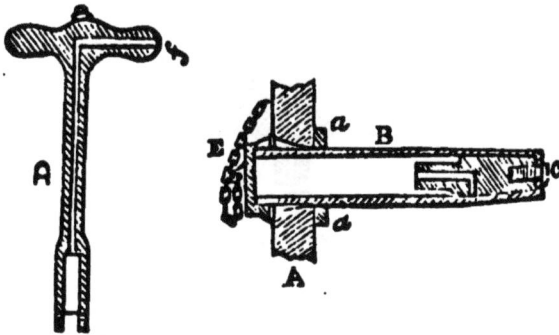

Fig. 14.

the tube B of the cock is fitted, and is held tight in its place by a nut *a a*, screwed on withinside. At the other end of the tube B, a plug C is fitted, by grinding it into a cone, and it is kept in by a screw. This plug has a hole up the centre of it, and from this a hole proceeds sidewise, and corresponds with a hole made through the side of the tube when the cock is open; but when the plug *c* is turned round, the hole will not coincide, and then the cock will be shut. D is the handle or key of the cock, by which its plug is turned to open or shut it: this handle is put up the bore of the tube, (the cover E being first unscrewed and removed,) and the end of it is adapted to fit the end of the plug of the cock. The handle has a tube or passage bored up it, to convey the beer away from the cock when it is opened, and from this a passage *f*, through the handle, leads, to draw the beer into a glass or tumbler. The hole in the side of the plug is so arranged, that, when the handle is turned into a perpendicular direction, with the passage *f* downward, the cock will be open. The intention of this contrivance is that there shall be no con-

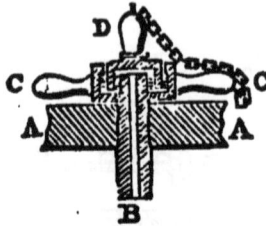

Fig. 15.

siderable projection beyond the surface of the tun; because it sometimes happens that a great hoop of the tun breaks, and, falling down, its great weight would strike out any cock which had a projection; and if this happened in the night, much beer might be lost before it was discovered. The cock above described, being almost wholly withinside, and having scarcely any projection beyond the outside surface of the tun, is secure from this accident.

Fig. 15 is a small contrivance of a vent-peg, to be screwed into the head of a common cask when the beer is to be drawn off from it, and it is necessary to admit some air to allow the beer to flow. A A represents a portion of the head of the cask into which the tube B is screwed. The top of this tube is surrounded by a small cup, from which project the two small handles C, C, by which the peg is turned round to screw it into the cask. The cup round the other part of the tube is filled with water; into this a small cup D is inverted; in consequence, the air can gain admission into the cask when the pressure within is so far diminished that the air will bubble up through the water, and enter beneath the small cup D.

PRACTICAL PROCESSES OF BREWING.

UNDER this head we will speak of the different processes of brewing, as practised in different countries; of the process of brewing porter on a small scale; of the most approved method of malting and brewing, brewing English home-brewed ale, &c. &c.

The difference of the English and Scotch methods of ale-brewing consists—1. In mashing. The English brewers run the water, at a temperature of 170° Fahrenheit, through the malt from the bottom of the mash-tun, and stir each mash. The Scotch brewers, on the contrary, fill the mash-tun with a sufficient quantity of water for the first mash, at a temperature of 175°, into which they run down the malt: they stir the first mash, preferring the use of oars to the mashing-machine, which latter they think taints the wort. The succeeding mash is effected by sparging or sprinkling at 180°, which commences at the time the first mash is drawn by the taps; and this sparging goes on constantly until they obtain the full quantity of worts required for the brewing. In boiling the worts, the English brewers, having drawn a sufficient quantity of wort from the mash, boil down to strength generally about two or two and a half hours. The Scotch draw a shorter quantity of worts, and boil down to strength generally in one hour and a half. 2. In fermenting the worts the English pitch at a high temperature, from 62° to 65°, and bring the gyle forward to cleanse into barrels within forty-eight hours. The Scotch ferment at a low temperature, from 50° to 55°, and work the gyle from eight to ten days, beating in the head of yeast occasionally, until the attenuation is judged completed, when they run the ale from beneath the yeast into barrels, where no more fermentation takes place. English brewers, in cleansing, mix the yeast on the head of the gyle with the wort, and run the whole brewing into barrels to cleanse: this is accomplished by keeping them regularly filled up until the fermentation ceases, and as much yeast as possible separated from the ale. The Scotch cleanse in the gyle, as already described. In both methods of fermentation, which may be distinguished by the quick and slow method, the judicious brewers of both countries never wish to carry the degree of heat more than ten degrees higher than that at which the wort has been set to fermentation.

Thus it will at once be observed that, to give a practical detail of both processes of brewing, the object will be much better accomplished by carrying through a description of a single brewing of ale by each of these methods. It may not be amiss, before doing so, to take some notice of matters connected with these two modes of brewing, and of such previous preparing and judging of materials as may be thought conducive toward producing ales of the finest flavour and quality. As we see there is so much difference existing in the methods of brewing Scotch and English ales, by which mode can that of the finest quality be made? It is a hard question to answer, and is of no little importance. With common brewers, local prejudice and taste must be so much studied, that in most places it is not safe for them to change their established system; but in large towns this may be got over, where the advantage of improvement is so decided, and the sale quick, that the change in the system of brewing cannot affect the regular production sent off.

Brewers who adopt the Scotch system under such circumstances, may find it very advantageous; but to small brewers, who do not brew but a few barrels three or four times a week, the adoption of the Scotch method would not be so beneficial.

The difference in fermentation constitutes the principal difference in the two systems. In many parts of this country as well as in England, brewers mash two or three times weekly, the whole year round. Were such to adopt the Scotch mode of brewing, they must have eight or nine gyles constantly in operation, which would be rather expensive. Where there is such a large stock on hand, it is liable to run into the second fermentation, in the summer months; and this forms a great objection to it. There are parts, however, of the Edinburgh mode of brewing ale which may be profitably adopted in most breweries, without in any degree deranging their usual methods of working; for instance, the method of sparging the mash, and in shortening the time of boiling the worts: a judicious application of these parts of the Scotch method to the quick system of fermentation, would be of importance both toward economy and improving the quality of the ale.

When the ale is intended to be bottled and kept for a length of time without adding more hops, the Edinburgh system is certainly the best to be adopted; but when intended for draught, and when the run is constant, and the demand instantly to be supplied from the brewery all the year, the English system is greatly preferable; so much so, indeed, that some of the Edinburgh brewers have rather followed the English method in brewing their ales to be drawn from the butt, and have found the alteration of much advantage. Therefore it will be admitted that both methods possess superior points to each other, and that to a brewer who studies improvement, a knowledge of the two systems in practice may become valuable, according to the circumstances in which he may be placed.

Ale which is brewed by private families in some parts of England, is the best to be found in Great Britain. With such, home-brewed ale is made and kept in a state of perfection which ale of no other country has excelled.

If the best malt and hops are selected, the first mash drawn of sufficient strength, and the second mash so regulated as to make up the quantity of worts required for the brewing, the boiling of the worts with the hops, only continued so long as necessary to extract their aromatic bitter, and the fermentation managed with judgment, ale is produced approaching to wine in quality, which may be kept in fine condition and pure flavour for years. This home-brewed ale is always made by the process of stirring the mashes, and of quick fermentation, so that it is quite evident that the superiority of one ale to another is not to be attributed either to mashing or fermentation. Both in England and Scotland, ales of an inferior quality are produced even when the best materials are within reach of the brewer. If fine hops and malt are employed, this should never happen.

Yet it is undeniable that a marked difference exists in the quality of ale made in the same localities, where the brewers use equal quantities of malt and hops, the same kind of water, and work out the process by similar methods.

In malting and brewing, there are a few points which should be brought to bear, as much depends on them in the production of good ale:

1. In malting, after the process of artificial germination has been carried to that particular stage sufficient to secure the greatest outcome of starch, to kiln-dry the malt by regularly increasing heat, from 100° to 150°. If the kiln-heat is too low at first, the malt renews its growth and begins to spring; when too high, it will blow or expand, and, when the heat is then rapidly increased, get scorched. A great deal depends on kiln-drying; and brewers who have maltings cannot give too much attention to this branch of their business; nor to the next point, which is to regulate their making of malt by the quantity they employ in brewing, so as to contrive always to work with as fresh-made malt as possible. With fine, recent-made malt, hops of good quality, and common care and skill in conducting the process of brewing, ales of the richest flavour are sure to be the result. 2. In mashing, to regulate the mash by heats suitable to the age and quality of the malt in operation, and the state to which it is crushed or ground. 3. The second mash is never to be drawn to a greater length than can be boiled down to strength within that particular time necessary to cut the worts and secure to the brewing the fine aromatic bitter principle of the hops. 4. If fermenting by the slow process, to regulate the heat of the gyle with such a precision that the attenuation and increasing heat of the worts should be commensurate, slow, and progressive.

By the use of the refrigeratory tube or worm this is accomplished, and with which every gyle ought to be fitted up; and if by the quick method, to watch the favourable time for beating in the yeast; the cleansing not to be hurried, but the worts allowed full time to strengthen the yeast before it slackens; and in both methods to work the gyle so as not to allow the worts to rise more than 10° or 11° of heat higher than when first pitched. 5. Cleansing after the fermentation of the worts, either by the quick or slow method. Thus, attenuation, or, in other words, the partial resolution of the starch-sugar of the malt into alcohol, and the preparation or cleansing of the yeast from the liquor, are all processes which require the brewer's skill and experience to carry them through properly. The ale is cleansed in the gyle by the slow method of fermentation; and when all is right, and the attenuation brought down to the point desired, it is run into the same casks which are sent out to the customer.

With the Edinburgh brewers, little or no fermentation takes place, and ale is never racked into other casks; but in Alloa, Stirling, and Perth, which are the best districts for brewing ale next to Edinburgh, they run the finished ale into butts, and afterward *rack* into barrels, as orders are executed. These two methods of cleansing the gyle, in the Scotch system of brewing, are particularly worthy of the notice of the reader: both methods are the best suitable to the respective localities. The Edinburgh brewers pursue their method because their ale is sent out at once to the customers' cellars. The Alloa district brew large quantities which are sent to Glasgow, and other parts, generally for immediate use. In *racking*, the Alloa brewers prepare what they term fillings, which are worts of the same brewing, set at the quick fermentation heat, 60° or 62°, and use part of this store in racking, putting an English pint into each barrel. The Edinburgh brewers rely on the fine condition of their ale, and add nothing whatever before sending out the stock. It has been necessary to be particular, that brewers may understand precisely the two methods of finishing the gyle by the Scotch system.

Sometimes brewers overturn it into a clean tun or square, to check too rapid fermentation and cool the whole brewing, or to prepare for keeping a length of time; but for immediate consumption, the two methods above described are adopted.

When cleansing by the English method, or quick fermentation, one point may be particularly brought to notice: that is, when the working of the worts in the barrels has nearly ceased, to fill and keep them filled up with ale of the same brewing, from a barrel which has been regularly worked forward with the rest, and not to draw the trough or plus-tub too close, for the sake of getting as much wort scraped together as to finish the filling of the barrels. Some brewers pay attention to this necessary precaution, but others do not; and whatever is useful in brewery practice should be put on record.

In such a complicated process as brewing, of course there are many different opinions as to best methods of carrying on the operation, but the whole at once resolves into this:—From a given quantity of malt and hops, required the greatest possible quantity of ale of equal strength and of the finest flavour. One man may produce twenty hogsheads of excellent ale from a given quantity of malt and hops, and be esteemed accordingly; but he who can produce twenty-one hogsheads of as good ale from the same quantity and quality of materials, must be regarded as much more successful.

Since the beginning of the present century, great progress has been made by practical chemists, who have turned their attention to investigating this very difficult art, both in Europe and in this country. Their valuable researches have placed it on the rational foundation of science. Theories of the whole process are very nearly established that set every thing in its true light; and practical brewers may, with common industry and research, very readily give a good reason for every part of the process. The art of varying the operations of brewing, so as to produce ale of a distinctive character, such as Edinburgh or English home-brewed, or common ale, must be of advantage to every practical brewer.

In the course of the present remarks, time will not be taken up in describing the utensils used in brewing, as they have already been noticed. It may be simply stated, however, that the copper is built so high as to command all the other utensils of the brewery. The mash-tun is placed as near it as convenient, and beneath is the underback, large enough to contain the whole worts of the mash. In every brewery of magnitude, there are several boilers for boiling the worts together, or separately, and preparing water for all other purposes. If all the utensils are properly placed, and are in good proportion to each other, it is almost unnecessary to say that the working of the brewery is rendered more easy and economical, and the various processes are carried on with greater ease and satisfaction.

The capacity or size of the utensils should always exceed the calculated quantity of the beer they are intended to make. A boiler to brew twenty barrels of ale should be large enough to hold from thirty-five to forty barrels of water: it requires all that capacity to hold the worts drawn from the mash, and which must be so reduced by evaporation, in boiling and cooling down to strength. The boilers employed to boil the worts should regulate the size of all the other utensils.

Sometimes fanners are placed over the coolers, to assist in cooling the wort. Should the roof of the cooling-room be low, or the situation confined, fanners are of course advantageous: but where it is high, and the apartment quite large and roomy, with the windows sufficiently open, they are prejudicial. It is to supply a deficiency in the construction of the cooling-room that they are used, and nothing more, because they disturb the worts in the process of depositing the coagulated fecula and other vegetable sediment, which, when retained, helps on the acetous fermentation afterward. Some intelligent brewers, it is well known, are of the opinion that the worts get rid of this sediment or precipitated fecula during the process of vinous fermentation, and adduce the practice of distillers, who run these vegetable remains into their gyles to increase their attenuation; but it has always been found that the purer the wort can be got cooled down and pitched, the better and purer has been the ale; and that where fanners were used, the damage they occasioned was greater than the risk of time lost in cooling by atmospheric influence alone.

Malting is a process which is well known and has been often described ; but it may be remarked, that as the artificial germination of barley, and kiln-drying to stop that growth, which makes it into malt, is often part of the brewer's business, he ought, at the commencement of each brewing, to be a judge of the malt in operation to regulate his future proceeding.

Some writers have stated that the hordein or starch of barley is converted into sugar by the process of malting ; this does not stand the test of more recent investigation : neither is it correct to say that this process either destroys vegetable life or effects a change in its constituent principles ; the only change that takes place is, that barley, during the process of artificial germination, loses part of its gluten and mucilage, which are taken up by the rootlets and acrospire, with a small portion of sugar, which is developed at the same time as nourishment for these parts of the corn.

When malt is crushed and coarsely ground, and treated with twice its weight of water at a temperature of 160° to 180°, and allowed to digest or infuse for two or three hours, it is converted into glucosin or starch-sugar, and held in solution. This is simply the process of mashing to make strong ale. Therefore, in preparing the malt to undergo this process, the brewer must judge from its appearance in colour and quality how far it may be most profitable to grind it rough or small, to obtain the strongest wort. During the course of the year, he may have various qualities to operate with, and it requires skill and judgment to manage this part of his business. Malt recently made requires care in grinding. If it is of fine quality, the rollers of the mill should be slackened a little, not to crush it too fine. Malt of this description always yields worts of the richest flavour.

When old, pale malt ought never to be ground small, nor mashed at a high temperature. It is liable partially to set, and yield a turbid wort. Brown and amber malt bear a high temperature in mashing—185° to 190°; but it is neither safe nor prudent to go beyond 185° with any malt whatever in the mash. 175° is the best and safest heat, with average malt of every description, that can be used. High heats in mashing certainly produce the strongest, but not the finest worts. If ales are brewed by mashing at a high temperature of water, vegetable extract comes over with the starch-sugar, and, not being got rid of during the succeeding processes of boiling and fermentation, remains in solution in the finished ale, and disposes it prematurely to assume the acetous state. Experience must be the guide, in all cases where the brewer has control of the matter.

Every brewer is to some extent the judge of what quantity of hops are to be used in making beer. But the use of them in boiling the wort, the nature of the bitters they impart, and the best method of extracting the aroma of the plant, are very different matters, and merit particular investigation by those whose aim it is to make the very finest quality of beer.

It is not only necessary to know how to select hops, but it is as requisite to know how to preserve them in a proper state. Fine hops are sometimes bought, and, on arrival at the brewery, conveyed to the loft, and kept until worked up. If kept in stock for any great length of time, they

get winded, and lose all their most valuable properties.
When hops first arrive at the brewery, each pocket or bag
should be subjected to the screw-frame pressure, which com-
presses it to two-thirds of its size; it is then corded, and
the hops will keep all the season sound and fresh. For
the methods of extracting the aroma and bitter principles,
the reader is referred to the proper place in the order of
brewing.

The fine condition and keeping quality of malted liquors
to some extent depends on the water with which they are
made. Brewers who are scantily supplied, or have occasion, as scarcity or drought occur, to use water inferior to
that which they commonly work with, cannot be too much
on their guard in this respect. In some instances it has
been observed that brewers and victuallers, both in Scotland and England, use brackish water supplied by wells
on their own premises, or from some neighbouring rivu-
let, which contains too much vegetable remains,—in both
cases injuring the fermentation and damaging their ales,—
where springs of pure water were not far distant. The
expense of bringing such into the work, must, in these
cases, be compared with the prospective benefit to be de-
rived from their use, and guide decision; for, both in
malting and brewing, it is of the greatest importance to
possess a constant supply of the best water. The best is
pure, soft spring water, such as rises from chalk or lime-
stone formations. River water that flows in a hilly district is also generally good: when its course is through
moss or level districts of country, it holds vegetable remains in solution, and the yeast takes a character from its
use as well as the extract from the malt. Ale made with

it is always soft, rarely in very good condition, and apt to be readily influenced by atmospheric change and run into acidity.

Every brewer has it in his power to make a selection of materials. The preparatory steps to be taken in bringing them into operation to the best advantage, and in carrying through the various processes with skill and economy, so as to secure a production of the best possible quality commensurate with the quantity of material employed, must now engage his attention.

In the six following divisions, the whole art of brewing is comprehended :—1. Malting, and preparation of malt; 2. Mashing; 3. Boiling; 4. Cooling; 5. Fermentation; 6. Cleansing. Storing and management of vats, and racking and mixing of stock for delivery, may be reckoned another division; but the six enumerated combine the art of making ale and bringing it forward to its first finished state.

To be competent in his business, a brewer must examine each of these processes by itself; he must study each in all its bearings, and endeavour to find out a reason, founded on sound principles, for the action of the matter before him, for the cause of that action, and for the successive changes which are produced on the malt and water by the energy of caloric, as he combines one part of the process with the next, until the whole experiment is successively carried through to the desired result.

When pale malt is dried at a high temperature, it keeps a much longer time, without imbibing so much moisture as to injure the future extract, than when dried at the heat which maltsters generally employ; and it will always be found that when dried by the law of Dr. Thomson, the saccharine extract will not only be of greater weight, but of finer flavour, than when dried at a low temperature, and preserve its qualities for a much longer time when kept in stock. In kiln-drying malt, however, at a high temperature, very great care is necessary to preserve the colour and mellowness.

Always bear in mind that it must be gradually brought up from 90° or 100°,—with this especial observance, that the kiln requires to be previously tempered to these heats before the malt is spread on its floor. The heat is gradually raised to 120° when the malt has been on the kiln two or three hours after it is turned, but not before. During this increase of heat it is turned again, and the temperature gradually raised to 140° and 150°; now precisely at this period it may be continued and finished on the kiln, as is the present general practice, or the heat may be increased further up to 170° or 175°, according to Dr. Thomson's law, the maltster watching its colour and trying its condition. The malt must be pale and mellow, the ends of the grain not having the least appearance of being dried brown, or scorched, and having preserved this condition at these increased heats, the furnace or fire-grate is drawn, and the kiln is finished, by allowing the malt to remain on the floor until the comings (rootlets) are trodden or screened off before removing it into stock.

SCOTCH SYSTEM OF BREWING ALE.

THE system of brewing ale adopted in Scotland exhibits superiority only in those places where justice is done in taking the previous steps of malting the finest quality of barley; in providing hops of that particular description necessary for imparting a fine aromatic flavour, and in carrying through the whole process with that care and skill which experience in making a fine quality of ale bestows. In these respects, Edinburgh, or rather the Mid-Lothian district of Scotland, maintains a decided eminence in the production of ale. The remarkable and uniform good quality of these ales gives the brewers unity of character, similar to those of London in the manufacture of porter.

However, there are other places, both in Europe and in this country, in which ales are made of a very fine description. The Alloa and Stirling districts, including Perth eastward to Montrose, have been celebrated many years, and may justly rank with Edinburgh in their productions. The methods of working are nearly similar in all these places, differing, however, in some parts of the process so as to give the ale a character of its own, which is easily distinguished in any part of the country where it is presented for use.

In giving a detail of the methods adopted in working out the Scottish system, such observations will be offered as may afford the practical brewer useful information in the various processes of mashing, boiling, cooling, fer-

mentation, and cleansing, compared with the English method of these operations, so far as may give an extended
view of the whole art of brewing.

1. MASHING.—The theory of converting the starch
of barley into sugar, or, as generally called, *glucosin*, has
not yet been proved by the demonstration of experiment,
though many practical chemists have turned their attention to it.   For a long time it has been known that when
starch is boiled four or five hours with dilute sulphuric
acid, the formation of glucosin, or starch-sugar, is the result;
and that when barley, either in the state of raw grain or
artificially germinated, is dried in a kiln at a temperature
gradually increased from 100° to 150° or 170°, and, there-
after crushed or ground, and treated with water from 150°
to 180°, the starch it contains is also changed into glucosin, or sugar, similar in its constituent principles to that
obtained from common starch boiled with dilute sulphuric
acid.   No acid is known to exist by which the change
of the starch of malted barley is converted into sugar.
Therefore it must be held, either that the agency of water
heated up to 170° affects the change, by a portion of the
water combining with the starch, or that the barley acquires in kiln-drying a new property, by the action of caloric, sufficient to convert the starch into sugar.   In this
last case, in the absence of demonstration by experiment,
recourse must be had to hypothetical reasoning.   Some
have urged, that when barley, malted or unmalted, is
dried on the kiln, it loses a fifth part of its original weight,
by the water it contains being driven off by heat, or, in
other words, that it parts with oxygen and hydrogen in
such proportions as form water.   In kiln-drying malt, the

continued action of heat on its substance might produce carbon, which, uniting with the oxygen, would form carbonic acid.

Let it be supposed that a brewing of twenty barrels of ale is required to be made from eighty bushels of malt and eighty pounds of hops. The utensils being all in a state of readiness, and the malt and liquor (water) prepared, three or four barrels of liquor, at a temperature of 180°, are first let down into the mash-tun, and, at the same time, the sluice of the malt-bin is opened, and the malt and remainder of the liquor, at 175°, run down together, and are immediately mixed and stirred by men with oars. About three quarters of an hour is occupied in mashing, according to the quantity of malt in operation. The whole quantity of liquor for the mash is twenty barrels. The head of the mash-tun is now closely covered, and the mash allowed three hours to extract.

When the liquor is let down into the mash-tun the heat is 175°, that of the grist 55°. Ten minutes after the whole is run down, the heat of the mash at the surface is 138°. The water has lost 37°, but the malt has gained 83° of heat. The weight of the water was two and a quarter times that of the malt: the mean heat of the mash, therefore, should have been 142°. The heat which, contrary to the usual law, is least at the surface of the mash must account for the deficiency in any trial the brewer makes of heats of the worts after the mash is laid on.

The mash having remained three hours to extract, the head of the tun is uncovered, and the sparger fixed. The sparger is a cylinder made of copper or other metal, about five or six inches in diameter, but, of course, made in proportion to the size of the mash-tun. It is closed at both ends, and nearly so to within a foot of the centre, which is open with a cross division, against which a run of liquor by a spout from the copper, strikes and sends it round the tun.

Across the latter an iron bar is fixed, on which the sparger is placed on a pivot. Its two arms extend the width of the tun: the underside of these are pierced with small holes similar to the mouth of a watering-pan, from which, as it revolves, the liquor escapes and sprinkles the mash.

The liquor in the boiler being tempered to the heat required for sparging, (sprinkling,) 185°, the tops of the mash-tun are slacked, (set,) and the worts are permitted to flow slowly at first, until they become transparent. At the same time the sparger is put in motion with as much liquor at 185°, as of worts which flow from the taps, care being taken that the head of the mash is never dry; and this flow from the taps, and, sprinkling on the surface, go on until the required quantity of worts for the brewing is obtained.

When the taps are set, the heat of the mash is 140° at the surface, and 150° at the bottom of the mash-tun. These are generally the degrees of heat in the Edinburgh method of mashing, and they are equivalent to the heats in the English method of mashing and stirring with machinery. The law which regulates the cooling of fluids is reversed in the worts of the brewers' mash-tun, the heat being the greatest at the bottom, instead of the surface. This arises from the malt settling down when left in a state of repose, and preventing the colder stratum of wort descending from the surface and displacing that which is beneath. Therefore the worts at the under part of the mash-tun are confined, and cannot ascend. The heat lost escapes through the side of the mash-tun, but this is in a very trifling degree. The pure worts, when drawn either from the taps or from the surface, immediately obey the general law.

In sparging, when the malt swells up a little, it is a favourable sign of the extract being good. It has been already observed that the head of the mash should not be run dry in sparging. The reason is, that when such is the case the goods sink, and the surface cracks: the liquor then percolates without extracting its due share of the saccharine matter from the malt.

Should the sparging-liquor be at rather a high temperature, there is another evil: it dissolves the starch of the malt without converting it into sugar, and escapes with it by too rapid descent through the grains, rendering the worts opake, and endangering the future quality of the ale.

The method of laying on the first mash having been described, together with the process of sparging, the strength of the first wort is ascertained by the saccharometer to be 96 pounds per barrel. When two or three barrels of the worts are drawn, the weight should be ascertained, as they are transparent, and show their real strength and quality, on which the brewer depends to regulate his future proceeding. The weight of the worts, when the mash is in the process of transfusion by the sparging-liquor, must be carefully watched. As the process advances, the strength of the sparges diminishes gradually until the weight is down to from 10 pounds to 15 pounds per barrel. In the present case, the whole wort required for the brewing is 30 barrels, which, being recovered and pumped up into the wort copper, is found to be of the weight of 72 pounds saccharine extract per barrel.

It will be observed that sparging or sprinkling is merely a continuation of the first mash, and that the difference of extracting the saccharine matter by this method and that of stirring the second mash by the English mode, and allowing it time to infuse, lies in drawing off all the extract the brewer desires, by the continuous sprinkling on the surface, until the whole quantity is obtained; while the English brewer has to lay on a third mash for the same purpose, at the disadvantage of being obliged to boil down a longer time to strength. A judicious application of the sprinkling method in the English system would be highly advantageous in obtaining the full saccharine extract with a shorter quantity of wort, and thus shorten the time of boiling, the prolongation of which is so injurious to the quality of the ale. If we suppose that the first two mashes were laid on and drawn at a fourth less quantity, and that, in taking off the second mash, the sparger was applied and slowly wrought, there cannot be a doubt that a stronger wort would be obtained, so as to shorten the boiling to an hour and a half, and thus preserve a considerable portion of the wort lost by evaporation, and the fine aroma of the hops besides. The Scotch neither get a stronger nor a finer wort by sparging, than the English brewer by stirring the mashes and giving them time to infuse: on the contrary, the former, by sparging, send down a considerable portion of the small dreg, or dissolved starch, without being converted into sugar, which descends through the grains, and escapes into the underback, rendering the wort opake, and not unfrequently carrying a cloudiness into the finished ale.

The brewers of Edinburgh are well aware of this : the superior quality of material they employ, and the great care and skill with which all their operations are conducted, enable them to regulate their first mash with the sparging process, so as to afford them all the advantage the English derive from stirring and infusing the second and third mash, without any necessity for damaging the wort by long boiling.

The first mash regulates and gives a character to the whole brewing, in both processes; the second, whether by sparging or stirring, secures it with regard to strength and flavour, and is hardly of less consequence than the first.

It is the opinion of many very intelligent brewers, that when the malt is first placed in the mash-tun, and the water run down into it through the bottom, 170° is the best and safest heat that can be used. 180° is the best heat when the water is first run into the mash-tun, and the malt shot into it from the hopper above. From these different methods of working the first mash, these heats are equivalent; and 175°, being the mean, may be taken as the best that can be employed. We find that this is precisely the degree of heat at which, according to Dr. Thomson, malt had preserved its virtues in kiln-drying. The coincidence is remarkable, and leads to the conclusion that, in proportion to the temperature at which pale malt has been finished on the kiln, the heats in mashing ought to be regulated to produce the greatest saccharine extract.

The quantity of liquor (water) required for the first mash, in the Scotch system of brewing, cannot be determined by any arbitrary rule. One and a half to two barrels of water for every quarter of malt in operation, are used by brewers, according to their views and future disposal of the ale. Those of Edinburgh prefer a rich, strong extract from the first mash; and as, by the process of sparging, they can transfuse the second, or rather the continuation of the first mash, to the required length, they always keep within two barrels of liquor to each quarter of malt.

It is necessary, to form a fair estimate of the advantage of the two methods, as practised in England and Scotland, to have regard to the views of the brewer in the disposal and future consumption of the ale. Until within a few years back, in Edinburgh it was generally made to be sent out to the publicans and grocers for the purpose of being bottled. The methods of mashing, boiling the hop-worts, and fermentation were the best that could possibly be adopted for brewing both their October or winter stock, and their summer or keeping ale. As the price is fixed according to the strength of the wort, and rated at a certain price per hogshead, the strength and flavour of these different priced ales required to be as equal and uniform as the brewer could possibly preserve; and thus the system of brewing became fitted for the production of such a quantity of malted liquor.

The general consumption of beer and ale, in England, is in draught from the cask; and the English system of brewing is as admirably adapted for the purpose as that of Edinburgh is for the consumption from bottle.

It is for the brewery proprietor to judge, in effecting a change in his system of brewing, first, whether or not such parts of the process could be adopted with advantage, without the great risk and expense of altering his utensils and general arrangements.   There is no difficulty whatever in the matter.   An English or American brewer, without any alteration of the utensils, may adopt, at any time, the Scotch modes of mashing and boiling the worts; and, according as these are judiciously carried through, it would be certainly attended with very great advantage, keeping the strength of the ales out of view altogether; because just as good ale can be brewed by one system as by the other.

Bear in mind that economy in short boiling, and in obtaining a fine aromatic extract from the hops, are valuable considerations, and are of such easy attainment, that the subject must certainly attract the attention of brewers.

Every brewer should study well the process of mashing. Here it is that the formation of glucosin, or starch-sugar, gives a character to the brewing.   In the malt there is an aroma as well as in the hops, which requires to be as carefully preserved in mashing as that of the latter in boiling the worts.   It is only by extracting and preserving these virtues from the materials in the greatest degree, that malted liquors in the highest state of perfection can be produced; and it is on this, and by economy in conducting each process, that many improvements in brewing depend.   The best temperature of the water for mashing has been repeatedly noticed, as also the quantity to be used in proportion to the quantity of malt in operation, with the time and method of working; but a very important point is now presented to the brewer's consideration, and that is, from the state of the worts, as obtained from the mash, to determine his future proceeding.

When all the worts have been secured from the mash-tun, and being in possession of a certain quantity of proved strength, he should be able to calculate the evaporation that must take place during the process of boiling and cooling. Not only should he know the amount of loss by evaporation from the boiler and coolers, but the destruction of saccharum which takes place during these operations, and thus be enabled to calculate the condensation or increase of weight down to the fermentation point.

If superior malt comes into operation, and yields an excess of saccharine extract, he confidently varies the future process to secure the advantage and cleanse a greater quantity of ale.

When the brewer is his own master, in these matters he acts at his own discretion. Should he be acting in the employment of another, his aim will be to acquire such practical knowledge, not merely to improve himself in his profession, but for the benefit of the establishment he is engaged with.

The arts of distillation and brewing are very easy matters in one point of view. The best malt-spirits that the country can produce may be found in the hands of the Highlanders; and there are hundreds in England who make home-brewed ale in richness and quality superior even to that brewed in Edinburgh: all these artists trouble themselves very little about specific gravity, sac-

charine extract, or any chemical knowledge whatever:
they have choice and plenty of material; but in regular
establishments, where certain quantities are required from
given proportions of materials, and where, indeed, from
competition and expenses of business, it is requisite to
work with the greatest economy throughout, it is fitting
that the person who has charge of the operative departments should have every facility of acquiring that general
knowledge which will be satisfactory to himself and profitable to his employer.

In proportion, generally, to the quality of the barley,
does malt differ in quality. The weight of saccharine
matter it contains, therefore, varies, according to quality,
from twenty pounds up to thirty pounds per bushel.
There are hundreds of brewers who have no choice whatever in selecting malt—they work up whatever comes to
hand.

A variety of malt is operated with in the course of the
year,—even made in their own malting, where good and
indifferent samples are taken off the kiln and injudiciously
mixed in stock. It is not every brewer that tests his
worts by the saccharometer,—many work by guess; the
same proportions of water and degrees of temperature are
applied to all malt whatever, and the usual quantity of
hops weighed for each brewing. The gauge of the mash-
tun is known by a mark, and the time of boiling regularly
kept. The worts are cooled down to the predetermined
heat, and the usual pailfuls of yeast from the onset added.
The fermentation runs its course. The ale is cleansed and
is really good; but it may, and often does happen, one
very important addition is wanting—fine malt has been

in operation, and the wort has been five or six pounds per barrel richer than usual. This would have made two barrels in addition, in a brewing of twenty-five barrels, had the brewer possessed the knowledge to have weighed the worts from the mash, and worked out the process accordingly.

In describing the process of boiling and cooling, the rate at which worts condense by evaporation from the boiler and coolers will be noticed, and to which the reader is referred.

These observations on the process of mashing may be closed by a remark on the influence of the temperature of the atmosphere on that operation. Malt generally preserves an equable temperature, ranging from 50° to 60°, according to the situation in which it is kept in stock; 55° may be taken as the heat of malt in good condition. When struck by the mashing-water at 175°, the water loses and the malt gains heat in proportion to the weight of the substances. The mash, during the three hours, gains more heat from the chemical action of the particles of matter in a state of decomposition and recomposition in the mash-tun. On these actions the temperature of the atmospheric air has little effect; so little, indeed, that there is no occasion, either in winter or summer, for changing the heats in mashing, care being taken that the water is let down in a close trough or spout, and that the head of the mash-tun is carefully covered up after stirring.

Two or three degrees may be allowed in sparging, when the thermometer is at the freezing point, or below 40°; but it is really of little moment—the chief danger in mashing is from too high temperature. The law by which fluids are cooled when exposed to the atmosphere, does not apply to the brewer's mash, as the heat there is greatest at the bottom of the tun, and least at the surface.

Much has been written on the danger from gluten and mucilage coming over in large quantities with the saccharine extract from the malt. These dangers are visionary, and cannot possibly exist. Barley, in a state of raw grain, does not hold more than eight per cent. of gluten, mucilage, and albumen, altogether; but when converted into malt, these substances almost disappear, being taken up by the rootlets and acrospire during the process of artificial germination. In mashing at too high a temperature, the danger is that the starch of the vegetable becomes mucilaginous, or *sets*, in place of being converted into sugar, a circumstance which strikes the brewer with more terror than when the false bottom of the mash-tun breaks loose and comes floating up in the mash.

2. BOILING THE WORT.—Having pumped up the thirty barrels of wort—seventy-two pounds of saccharine extract per barrel—into the boiler, and brought it through to boil for half an hour, forty pounds of the hops are added. Much care is requisite in boiling the hops, that they do not fry on the surface of the wort; by which is meant that they do not come to the surface by the boiling action and froth, and give out their aroma and bitter principle with the vapour that escapes, before being incorporated with the saccharine extract. This term is not acknowledged by the English brewers: in their method of brewing they get the bitter of the hops accumulating at the surface, and coming over with the boiling wort, when they use the open copper, by shutting the damper of the furnace, and beating down the head of the worts with an

oar : if an upperback is on the copper, such precautions
are unnecessary. Neither upperback nor double boiler is
used in Edinburgh. A machine is now employed, which
is inserted into the boiler, by which means the worts are
permitted to boil through, but the hops are kept beneath
the surface.

After boiling another half-hour, the remainder of the
hops is delivered into the boiler, and the boiling continued
half an hour longer. This is sufficient both to extract the
aroma and bitter principle of the hops, and to concentrate
the saccharine extract by evaporation down to the required
strength.

The worts in the boiler are eighty-four pounds per barrel, as indicated by the saccharometer, when cooled down
to 60°; but, in consequence of the evaporation which will
take place on the coolers, their gravity will be in proportion to the water driven off by evaporation; and as this is
an eighth part of their bulk, the increase of saccharine
extract per barrel will be in the same proportion, deducting the saccharum that escapes into the atmosphere during the process of cooling. The increase on the coolers
will be seven pounds per barrel on worts of eighty-four
pounds per barrel; their weight, therefore, when cooled
down to 60°, will be ninety-one or ninety-two pounds per
barrel. For in this brewing, calculating that three barrels
of wort evaporated, this leaves two hundred and fifty-two
pounds saccharine extract, and deducting one and a quarter pounds per bushel on the eighty bushels of malt in
operation, as the amount of saccharum destroyed, one
hundred and fifty-two pounds are left condensed in the

9

wort, which is about seven pounds per barrel of increase to the remainder of the worts on the coolers.

Like all the other processes in brewing, the process of boiling the worts requires much attention and care to bring it through successfully with the least possible loss, and to preserve the aroma and first bitter of the hops. In the boiling of worts it has been stated that the process combines four actions which affect them, viz. :—1. Boiling; 2. Evaporation; 3. Condensation; and, 4. The destruction or escape of part of the saccharum. The quantity of water evaporated is in proportion to the weight of saccharine matter condensed, adding to the amount of condensation one pound per bushel of the whole quantity in operation, which escapes in this proportion each hour the wort is exposed to the boiling temperature. The amount of condensation in boiling is one to ten per hour; that is, worts of the gravity of fifty pounds saccharum, strengthen five pounds per barrel in one hour's boiling; and, carrying on the same proportion, worts of one hundred pounds saccharine extract strengthen ten pounds per hour. Thus worts of one hundred pounds, in one hour and a half boiling, strengthen fifteen pounds; and when to this is added the increase of strength by condensation on the coolers, calculated in the same proportion as formerly, the weight of the brewing would be one hundred and twenty-four pounds saccharine extract per barrel, which is a sufficient approximation for practical purposes.

Thus is the brewer provided with the means of shaping his processes after the recovery of the worts from the mash, to make the most of the property intrusted to him, by working with economy of materials, and to produce the greatest quantity of finished ale from a known quantity of worts.

3. COOLING THE WORT.—According to the notion which primarily existed of cooling the worts after being finished in the boiler, the plan would be to spread them in large shallow vessels, and expose as large a surface as possible to the atmospheric influence, which, by evaporation, would cool them down in the quickest manner to the temperature desired for fermentation. It would thus be observed in ancient times, when the art of brewing malt liquors was first discovered, that otherwise the wort, by too long exposure, would spontaneously ferment and run into acidity. Throughout the long period since the art of brewing, and afterward, since the fifteenth century, when the art of distilling ardent spirits from wine and vegetable sugar were discovered, this plan seems the only one that ever has been followed, until within the present century, when distillers found out that by running the worts through pipes immersed in water, they could regulate the temperature to their desired point, and avoid the waste by evaporation altogether.

Having boiled the wort of the present brewing for one hour and a half, and allowed it to remain in the copper for the space of a quarter of an hour after drawing the furnace, their strength, as indicated by the saccharometer, being eighty-four pounds per barrel, it is then run into the hopback, and from thence spread on the coolers. When the worts are first run from the boiler the hops are stirred, to allow as much of them to escape as possible, and lodge at the bottom of the hop-drainer.

By pursuing this course, the worts will filtrate much

clearer into the coolers. The brewer always regulates
the time of the brewing so as to admit of the worts being
run into the coolers during the afternoon, that they may
catch the coolest time of the night to go down to the fermentation degree of heat required. The sooner they are
brought down to this point the better. Sometimes it happens, though rarely, that they *fox*, or set the backs, as it
is termed; which means that, in consequence of being too
long on the coolers, or from the latter not being perfectly
cleaned, they begin to ferment, and are apt to run into
the acetic state. They should, in all cases, be cooled
down and pitched to ferment within twelve hours from
the time of being run from the copper. As soon as the
worts are spread on the coolers, they immediately begin
to condense,—to concentrate in weight by evaporation,—
and to throw off part of the saccharine extract, all of which
goes on during the time they are exposed to the influence
of the atmosphere: therefore the process of cooling should
be studied by brewers, so as to carry it through with the
greatest possible economy and expedition, to preserve the
quality of the ale in operation.

A saccharometer is used to ascertain the amount of
condensation, which, if constructed on proper principles,
indicates, at any temperature of heat down to 60°, the
weight of saccharine extract the worts contain. The
quantity of wort evaporated may be calculated by the increased density of that which is left in the coolers, deducting the saccharine matter destroyed during the process.
The brewer may gauge the worts when they have cooled
down to 60°, and ascertain their quantity, as also their
strength, by the saccharometer. It is by comparing the

total quantity and weight with those of the mash, and striking the difference, that he will observe what has been lost during the process. This is of importance to the brewer, and he should be enabled to do so.

Every process should be carefully watched, and the worts tested by the saccharometer, from the time of laying on the mash until the worts are cooled down to the fermentation point. According to their quantity and strength, the process can be varied in perfect safety, so as to secure the greatest amount of finished ale. Without the utmost precaution—it cannot be too often repeated—waste occurs; and whatever may be the skill and experience of the operator, it will be readily admitted that economy is a qualification valuable in proportion to the judgment with which it is exercised. In brewing malted liquors, there is neither economy in sparing the materials, nor in using those of inferior description. The best malt and hops are the cheapest, in the end, that a brewer can use who studies his own interest; and it will ever be found that, in the manufacture of ales, their strength and richness of flavour will be something in proportion to the economy with which the process has been conducted.

The open cooler is the method in general use for cooling the worts, but there are various methods occasionally practised. Some distillers, it has been already noticed, run them through pipes of great length immersed in water, and thus cool them down to any required temperature; but this method does not answer brewers, who must get rid of the coagulated fecula and vegetable sediment, which distillers admit into their fermenting-tuns without hesitation, as these promote the fermenting process. A few new methods of cooling may not be unacceptable to the reader, and are worthy of the special attention of all practical brewers.

When the worts are first run from the copper, they are sent through a tube immersed in water, and lodged in a receiver which is capable of holding one or two hogsheads: after being subjected to the influence of the water a sufficient time, they are displaced by another charge, and the first spread on the cooler adjoining; and this process goes on until the whole contents of the copper are subjected to refrigeration. Part of the wort is saved by this process, which would otherwise be driven off by evaporation, and permits the deposit of the sediment on the coolers during the whole time they remain afterward until they go down to the fermentation heat.

Another method is to fix a flat water-tight receiver round the inside of the whole cooler, the bottom of which dips an inch into the wort. Cold water being run through this receiver, its influence strikes the worts beneath, and cools them down rapidly.

After being cooled in the usual manner to 75°, the worts are let down from the coolers, through a worm placed in a large cask filled with water, and are cooled down to the temperature the brewer requires.

The latter has occasionally been used in practice in summer brewing, and always found to answer the purpose. In all methods of cooling, where the evaporation of the worts is prevented, they must be first boiled down to strength in proportion.

An iron cooler has been adopted by some brewers, and

is also very useful, when, from the warmth of the atmosphere in summer, the worts are slow of going down to the temperature required. A quantity of water from the brewery-well is pumped into the cooler; the water is generally at 40° to 44°; the iron soon acquires this temperature, and, the water being let off and the cooler quickly mopped dry, the worts are run into it to the depth of an inch, and thus go down successively to the required heat, to be run into the gyle.

These methods are very useful. As they are at once simple and effective, they are worthy of consideration, as any brewer can adopt them with but little expense.

A regular method of cooling down the whole worts of a brewing, on a correct principle, to bring them to the temperature of fermentation, and at the same time to admit them to deposit their sediment, and flow pure into the gyle, and this on a scale of magnitude sufficient for large establishments, is still wanting.

This great improvement in the art of brewing will, no doubt, yet be effected. Were a utensil so constructed as to hold the entire quantity of the brewing, (or, when on a very large scale, it may be divided,) and a supply of water kept constantly running on the surface, this utensil at the same time being kept full from the hopback, which, on an improved construction, would require to be used both as a drainer and reservoir, the worts could be thus cooled down sooner than by the open method of evaporation; the action of refrigeration, being vertical, would carry down the sediment, and no loss of wort or of the saccharine extract whatever would be sustained. This method has been proved on a small scale, and there is no reason to doubt but that it would prove successful on the largest, and become a valuable acquisition to brewers and distillers. .

In cooling, it has been calculated that worts lose about one-eighth of their bulk by evaporation; but they do not acquire a density in proportion to the quantity driven off: a part of the saccharine extract escapes at the same time, which must be deducted, and the difference shows the increase of weight they acquire when cooled down to 60°.

There is an escape of about one and a half pounds per bushel of the saccharine extract on the whole malt in operation; but this is a mere rough calculation, which has been made by practical brewers, and requires confirmation by experiment. Sometimes it happens that the worts of the same brewing are boiled down in separate coppers, and that the hops are used in different proportions in the process, the weaker wort being reserved until the stronger is run through the hopback and the hops left in it.

The weaker wort is now run into the hops and carries off the strength which was left in them. The stronger and weaker worts are cooled down separately, and mixed in the gyle-tun in these cases.

Various are their constructions. Sometimes they are a square wooden frame, pierced full of small holes in the bottom and round all the sides. This kind is tedious to work, as the hops fill up the holes, and the worts come through slowly, but very pure. Another kind is close at the sides, and has open spars across the bottom, over which a hair-cloth is spread, through which the worts rapidly escape into the coolers. The best construction is that with a false bottom, like the mash-tun, with a stopcock to regulate the flow of the worts. This is the best kind in use, and ought to be adopted by all who would obtain a pure wort, and wish to manage the hops with the greatest economy.

The cooling of the worts, toward effecting the greatest possible saving, is of so much importance, that it will again be referred to when on the subject in the English method of brewing.

4. FERMENTATION.—After the worts have been cooled down to 53° and six galloons of yeast prepared, the first part of the process is to pitch the gyle; that is, to mix the onset of yeast and the worts together in the gylo-tun, to commence the process of fermentation, the weight of saccharine extract being 94 pounds per barrel. One barrel of wort is first run into the gyle, to which the six gallons of yeast are added and thoroughly mixed; the remainder of the worts are then pitched in full flow from the coolers, at the temperature, as already mentioned, of 53°.

Both in the slow and quick methods of fermentation, the quantity of yeast, and heat of the worts, must be varied a little according to the season of the year. In the slow method, one gallon for every four barrels of wort during the winter, and two-thirds of that quantity for the warmer spring and summer months, may be taken as the average quantities used.

The degree of heat of the worts at which the yeast is added is of the utmost importance, as it regulates the time of the process of fermentation.

According to the season of the year, and especially the existing state of the atmosphere, it ranges from 50° to 55° in the Scotch system. In the English system of quick fermentation, the range of heat is from 60° to 65°; in both cases being the best that can possibly be used for carrying through the respective processes, and obtaining the desired combination of alcohol and solution of starch-sugar to constitute strong ale.

These precise heats require to be completely understood, several writers having given latitude to a much larger range, which is apt to lead into error. When the heats are lower than 50° in the slow, and 60° in the quick methods, the fermentation is languid, and recourse must afterward be had to heat the worts in the gyle by artificial means. If above 55° or 65° in these methods, in the first instance the worts are apt to spring from the slow into the quick fermentation, and endanger the brewing; and if the heat is above 65° in the latter method, the fermentation runs too quickly up, and renders the ale liable to commence the acetic fermenting process.

For commencing fermentation, the mean heat was formerly stated as $52\frac{1}{2}°$ in the Scottish, and $62\frac{1}{2}°$ in the English system; and it cannot be too earnestly urged, that in both the chemical principle is the same, although the action differs in manner and time,—the result required being the resolution of part of the starch-sugar into alcohol in such proportion as to bring out the ale in the highest state of richness of flavour, and fit for keeping until required for use.

Another point also requires explanation. Store yeast for onset requires to be changed occasionally in both systems, or, to be more explicit, the process of fermentation requires to be commenced by a change of yeast from another brewery; for, when too long continued in use in the same brewery, it is found to work languidly, and become deficient in strength and quantity. It may be said to work in-and-in to weakness, until it loses the capacity of carrying over a due proportion of the sugar in a state partially decomposed, and thus loses the power of acting with energy when applied to fresh wort. In making this selection to commence a new brewing, the brewer should exercise the greatest scrutiny. It is necessary to know the kind of water which is used, as the yeast acquires a character from its quality, and affects another fermentation accordingly. In practice, it is found useful to have a change of onset every four months. It very much depends, however, on the care taken in keeping the yeast from one brewing to another as strong as possible.

Mr. Black, in his Treatise on Brewing, observes, that there is no occasion for a change in the fermenting principle at all, and that he never attempted to make it; but there cannot be any doubt whatever that it is requisite. The utility of the practice is universally acknowledged, care being ever had that the yeast is from ale made from worts as strong as that to which it is applied, and that it is as strong and fresh as can be procured.

For the first ten or twelve hours after the worts are pitched, and the yeast has struck, a very decided alteration takes place, and they are turbid and unsettled in appearance; and a scum of a grayish colour has gathered on the surface. In twelve hours more, a white circle, narrow and regular, appears round the edge of the gyle, and the surface begins to chip, and show irregular patches of white breaking through ; then these unite and shoot up in little pyramids,—a proof that the yeast is beginning to form on the surface, and that carbonic acid is escaping from the worts. This is the first stage of fermentation, which the brewer looks upon as an assurance that his gyle is in a healthy state. The whole head of the worts is now covered with froth, which the brewer watches, and, as soon as he judges that the yeast is sufficiently formed, the head on the surface of the worts is beat down, and the process of fermentation allowed to go on for twenty-four hours.

The Alloa district brewers have a method of quickening the fermentation at this stage of the process, which is very serviceable. They prepare a half-fermented wort which is termed *fillings*. Reserving half a hogshead from the coolers, they put these to quick fermentation at 62°, and by the second day of the gyle's age these fillings are ready. Ten or twelve Scotch pints—about five gallons English measure—are thrown into the gyle, the effect of which is to make the fermentation lively and healthful. These fillings serve another purpose, for which they are chiefly intended. By the Alloa method of fermentation, the contents of the gyle, when finished, are cleansed or run into butts, from which the ale is racked into casks as required, and the fillings are added to preserve its keeping quality.

The greatest care is exercised by the brewers of Edinburgh to extract the aroma, without allowing the bitter to be much infused, except in ale to be kept through the summer. When ale is exposed to heat, either in a warm apartment, or by a change from very cold to mild weather, the aroma of the hops held in it escapes, and, not having sufficient bitter for support, sometimes acquires a soft, weak taste. But brewers must study the public demand; and such occasional condition, even of the best kind, cannot be avoided.

Twenty-four hours after the head of yeast has been beat in, the renewed yeast comes thicker to the surface of the worts, of a light cream-colour, and of a firmer appearance. The progress of the heat and attenuation, or resolution of the starch-sugar into alcohol, must be carefully ascertained. The increase of heat altogether to the finishing of the ale, must not exceed 10° or 11°; but the attenuation required, being according to the future views of the brewer, cannot be fixed by any arbitrary rule. Ninety-four pounds of saccharine extract, in this instance, is the strength of the wort, and the attenuation required is that it shall be carried down to 45 pounds per barrel. Therefore, the duration of the process depends on regulating the heat until the attenuation is accomplished.

The heat should advance progressively, and is either kept in check or encouraged by the use of the tube which is fixed round the inside of the gyle, taking five or six turns from top to bottom, through which hot or cold water can be run at the pleasure of the brewer.

In eight days the heat has increased 10°, and the attenuation, as indicated by the saccharometer, is down to 50 pounds per barrel, the head of yeast on the worts having been plunged occasionally during that time.

For beating in the yeast there cannot be established any definite rule: sometimes it is requisite twice in one day, sometimes not for two days together: neither can time be fixed on to determine the duration of the process of fermentation. Much depends on the quality and quantity of yeast employed to commence the process, and the heat of the worts when set to ferment. The appearance of the gyle gives the brewer a good notion, during the process, of its healthy state; the head of yeast should have a broad rolling appearance, full to the sides of the gyle, and swelling a little to the centre.

Good yeast has a close texture, not glassy, nor studded with bubbles of carbonic acid, nor of a flat surface. Either in the slow or quick method of fermentation, when it assumes this appearance it is high time that it should be cleansed.

By close observation the brewer must determine when the gyle is ripe, and when all is well relative to heat and attenuation,—to cleanse. This is done in Edinburgh by running the clear ale from beneath the yeast into the same barrels in which it is sent out to customers. No further fermentation takes place sufficient to render it necessary to put the barrels on troughs: they are placed on open stillions, or on the floor of the cellar. The ale is cleansed into butts in Alloa, and afterward racked into casks to be sent out. Sometimes it happens that the gyle, in spite of the brewer's care, runs up to a high temperature, and the fermentation becomes rather unmanageable. In such a case, the contents of the gyle are run as clear as possible into a square or clean tun. The ale cools down a little, and, in twenty-four hours, it is racked into casks; but this method of tunning ought never to be had recourse to, except the state of the gyle requires it, as it flattens the ale and injures its quality.

The reader's attention must be called to the progressive state of the gyle during the process of fermentation, as explanatory, in some measure, of the different action which takes place in the quick and slow methods of working.

When the yeast first gathers on the head of the worts, in the slow method of fermentation, it gives out carbonic acid, and in proportion as it allows this to escape, and feels the influence of the atmosphere, it becomes viscid, and, were it not beat in, it would sink down through the wort, and leave it almost clear. It is beat down, before it approaches this state, into the worts. The principle of fermentation it still contains is thus mixed with the worts, and resumes its action. The more viscid part of the beaten down yeast disunites from that which holds the fermenting principle, and attaches itself to the bottom and sides of the gyle. As each successive formation of yeast comes to the surface, and is in turn beat in, the same process takes place. The viscid portion thickens at the bottom and round the sides of the gyle, until the alcohol begins to overpower the fermenting principle, and gradually would destroy it altogether. This is the time for checking the farther progress of fermentation by cleansing,— by which term the reader will perceive, that it means the separation of the ale from the yeast-formation in the gyle-tun.

On comparing the above description of the nature of fermentation in the Scotch system of brewing with the English method that will hereafter be described, it will be perceived that both these processes have one chemical principle in common, which, although differently wrought out, are equivalent in the production of malted liquor.

The English brewer pitches his worts at 62° with a large proportion of yeast. From these causes, they arrive at the full fermentation standard in thirty-six hours, and during that time run·up 10° to 12°, leaving the attenuation behind. The yeast is therefore beat into the wort, and mixed thoroughly together, and tunned immediately into barrels set on close troughs, to hold the worts that immediately flow into them.

As it is repeatedly filled up, the yeast begins to assume the viscid state, precisely on the same principle as that described in the Scottish process of slow fermentation; but, in place of lodging in the barrels, it settles down on the bottom and sides of the troughs, just in the same manner as in the Scottish gyle, until the formation of alcohol checks the fermentation and gradually stops it altogether.

The repeated overturning of the ale and filling up of the barrels, in the latter process, diminishes its temperature, and preserves it from acetic fermenting-heat, but it is better guarded against that danger by the formation of alcohol, which now goes rapidly forward until sufficiently powerful to arrest further fermentation.

It has been attempted, as shortly as possible, to give the reader a distinct notion of the nature of both these methods of fermentation; and it remains to say a few words of the proper degree to which ale should be brought down by attenuation, so as to preserve the richest flavour of malt and hop, and to afford the greatest satisfaction to the consumer.

It has always been the study of Scotch brewers to

effect the attenuation of the wort so as to combine the exact proportion of alcohol and sugar of the malt to constitute ale of the richest description; and although every part of the process of manufacturing malted liquor may be said to be of importance, the successful attenuation of the worts in the Scottish system of brewing ale may justly be deemed one of the most essential requisites to establish the character it has acquired.

This process must ever be a desirable object to brewers. When the resolution of the starch-sugar into alcohol is carried down to rather a greater length than necessary, it promotes, no doubt, the purity of the ale and its keeping quality, but it renders it too thin to the palate, and unm: s the nauseous hop-bitter, which always, more or l   comes over with the hop-extract. When strong wo   have not been sufficiently attenuated, on the con-ry, the ale has a sickly, luscious taste, and is apt to un into acidity. To avoid extremes, and to hit the exact proportion, therefore, requires all the skill of the brewer. The future disposal of the production ought to influence the process. Ales for bottling ought always to be atte-nuated lower than ale for draught from the casks. In making ales of this latter description, the brewers of Edinburgh have found it of advantage to keep them fuller of saccharine extract than they had formerly done. Ales for bottling would, in some instances, admit of the same improvement.

The term *attenuation*, as applicable both to distillers and brewers' worts, means the thinning or weakening of the saccharine extract during the process of fermenta-

10*

tion, by its resolution into alcohol. Distillers carry it as low as possible, to obtain the greatest quantity of the latter by distillation.

Its progress is checked by the brewers, as previously explained, to combine what they judge to be a good proportion of it with the starch-sugar undecomposed to constitute strong ale. However, the amount of saccharum left in the wort is merely the apparent weight: the real weight is concealed by the quantity of alcohol given off during the process, the specific gravity of which, below that of water, counteracts the weight of the saccharine matter in solution above it, in proportion to the alcohol formed from the starch-sugar held in the original wort.

The worts of brewers require a different management from that of distillers. For instance, when of the strength of one hundred and twenty pounds per barrel, on being put to ferment, the decomposition of the starch-sugar is very rapid at first; but in proportion as the alcohol is evolved the fermentation decreases, and would gradually cease altogether, were not the head of yeast beat in to renew the action, as in the Scottish system, or mixed in the gyle-tun, and run into casks to effect the same purpose, as in the English system of brewing; thus supporting the conclusion arrived at by Dr. Thomson, that sugar, in a state of partial decomposition, is the fermenting principle of yeast.

Let us suppose these worts of the weight of one hundred and twenty pounds saccharine extract attenuated down so low as forty pounds per barrel, their real weight would be sixty pounds, from the specific gravity of the alcohol evolved adding a proportional weight of saccharum in solution, which could not be detected by the saccharometer.

Therefore it should be held as a rule in brewing ale, that the stronger the worts the less is the proportional weight of alcohol produced, after a certain quantity has been evolved by fermentation sufficient to counteract its further production.

5. CLEANSING.—English brewers comprehend by the term cleansing, the mixing of the yeast and ale together in the gyle, tunning into barrels placed on close troughs, and continuing the fermentation until the yeast forms and separates from the ale; which is accomplished by keeping the barrels repeatedly filled up, until the fermentation ceases, and the process is finished. In the Scottish system, it is applied to running the ale from the gyle into the casks after it has been judged sufficiently attenuated, and leaving the yeast behind.

As previously stated, there are several modes of cleansing, which may be again noticed. 1st. The Edinburgh method, by which the ale is run, finished, from the gyle into the casks which are afterward to be sent out to customers. 2d. The brewers of the Alloa and Stirling district cleanse into butts, from which the ale is afterward racked into casks, an English pint of fillings or prepared wort being put at the same time into each. 3d. When the ale is to be made up for exportation, it is overturned into a square or vat capable of containing the whole brewing. It is allowed to remain in this twenty-four hours; fermentation proceeds a little, and attenuation takes place to the extent of one or two pounds per barrel. A decoction of hops is prepared, of a strength sufficient for the intended purpose, and this, with a proportion of store, is added when the ale is racked into casks. These additions are made to preserve it in its vinous state, calculated to keep it until its time of consumption.

When brewers overturn their ale into squares, and rack for home consumption, it is rather to remedy a defect than from choice. It may be advisable to use a square when the fermentation has run up rather high, and, if done judiciously, the ale may be brought out in a very good condition; but, after the gyle, all future racking flattens it, and injures the quality.

With the Edinburgh brewers, it is the practice to bring the gyle to the highest state of perfection, and tun into casks at once, where no more fermentation takes place. Neither fillings nor isinglass, in a state of finings, are used,—it being considered that when ale is finished in fine condition, these are not requisite.

The measure of ale in hogsheads differs considerably from the English standard. The Edinburgh hogshead contains sixty-three gallons, and the trade generally allow fifteen per cent. discount on settling accounts. The reader will take the difference of measure, which is nine gallons per hogshead above that of England, and the liberal money discount, into any calculation he may make, in forming an estimate of the comparative advantages of the English and Scottish methods of brewing.

Having brought forward to the process of cleansing, from the gyle-tun, the brewing of twenty barrels of ale from eighty bushels of malt and eighty pounds of hops, it is tunned into hogsheads and half-hogsheads. As soon as run from the gyle into the casks, it is considered finished; and, as has been already observed, no addition is made before being sent out to customers.

ENGLISH SYSTEM OF BREWING ALE.

In nearly all parts of England, the method of brewing ale is very nearly the same. The malt is ground immediately before the mash is laid on, and the water, generally heated to the boiling temperature, is allowed to cool down in the copper to the heat required for mashing. The heat of the first water is 170°, and 185° for the second mash: the common practice is to take off a third for small beer. It is necessary, to obtain as much of the saccharine extract as possible for the ale, to draw the first and second mash to such a length as to render it necessary to boil down to strength from two and a half to three hours. The hops are added when the copper comes through to boil, in most breweries, and are subjected to the whole time of boiling the worts. The worts are spread on the coolers, and being cooled down to the temperature of 60° to 65°, are pitched to quick fermentation, with about two gallons of yeast for every four barrels of wort. The fermentation in the gyle comes to maturity in thirty-six hours. As soon as the head of yeast is ripe, and begins to sink, the ale is tunned into barrels, the latter being first plunged through the wort; many brewers at the same time throwing in four pounds of flour and two pounds of salt for every twenty-five barrels in operation. The worts in the gyle rise in temperature from 10° to 12°, before being run into the barrels to cleanse. One hogshead of wort is reserved to fill up with, which, with that which comes over into the troughs in tunning, is sufficient to cleanse twenty-five barrels of ale. Within two days the process of this additional fermentation is finished, and the ale removed into stock.

The art of brewing malted liquor, in England, must be divided into distinct operative employments and classes, all of which must be examined separately before the adaptation of each to its particular purpose can be appreciated. The first class is, undoubtedly, the art of making home-brewed ale. This description of malt liquor is carried to great perfection by private families in every district of England. The ale made in England termed home-brewed is superior to the Scotch ale, for the reason that it is forced, by the quantity and quality of materials used during the process of brewing, into that pre-eminence, and by the management it afterward receives in the cellar, to maintain its flavour and acquire the quality of keeping,—both very important matters in the constitution of strong ale.

Another class comprehends the numerous body of common brewers,—and victuallers who brew their own ale are settled in every town and considerable village in England. A variety of causes operate to make a diversity of strength and flavour in the production of ale in every county, which arises from brewers in the long run coming under the obligation of studying the tastes and customs of those among whom they carry on their business. Climate, quality of malt and hops, water used in brewing, and method of working out the system, all combine to impart to ale a distinctive local character, and reasonably account for that variety in condition and flavour which is everywhere to be met with.

These causes, though, do not operate in the manufacture of the first class, or home-brewed ale, already mentioned. In every part of the country, from the nobleman's well-ordered and complete brew-house down to the cottager's economical contrivance, ale of the same rich flavour and strength can be made from the same quality and proportions of materials.

The method of conducting the process of malting differs considerably in England from that followed in Scotland, chiefly arising from the different qualities of the barley; the climate and soil of the former country being more favourable for its growth of a fine quality than those of the latter.

The following question, and one which is of no little importance, remains to be answered:—By which system of brewing can ale of the best quality be made from a given quantity of malt and hops, and to sell at the same price, to yield the same profit?

In order to judge properly, Edinburgh ale must be placed in competition with that made in any particular county in England noted for the goodness of its malt.

The making of ale is a distinct art from that of making porter. The methods of making ale by the quick and slow processes of fermentation, also render them so distinct as to require a different arrangement of utensils altogether. The making of home-brewed, as generally practised in England, is also distinct from Scottish ale, and from that of the common brewers,—the excellence of its quality generally depending on a sufficient quantity of malt and hops, and judicious treatment in the cellar, for its future preservation.

These are facts which no brewer can dispute. Therefore, to divide the business of brewing into distinct classes, and to investigate the process by which the malt liquors are produced, is absolutely necessary before a competent knowledge of the whole can be acquired.

MASHING.—We will now suppose that, at the commencement of the brewing, every thing requisite has been cared for,—that the utensils are all in perfect order,—that the material has been carefully measured,—and that the water in the boilers is sufficient for the purposes intended. There are eighty bushels of malt in operation, with eighty pounds of hops, from which it is required to make twenty barrels of ale. Every thing being in readiness, the water is let down to the mash-tun at the temperature of 170°, and ascends through the malt. By the gauge of the boiler, twenty-four barrels, the quantity required for the first mash, are let down; the machine is put in motion, and the mashing proceeds. It is kept going twenty minutes, which is sufficient time; when oars are employed, half an hour is requisite. The head of the tun being carefully covered, the mash is allowed three hours to extract. For the first mash, twenty-four barrels of water for ten quarters of malt would be considered extravagant by Scottish brewers; but it must be considered that, in the present operation, the worts require to be boiled one hour longer than by the Scottish method; and that, as a quarter of malt takes up twenty-eight gallons of wort, seventeen barrels of the latter will be requisite from the first mash, seven barrels being taken up by the malt.

The first mash having extracted for three hours, the tap is set, and the worts are permitted to flow into the underback.

The heat of the worts at the surface is 140°, and as they flow from the tap, 148° to 150°. They are run slow at first, to confine the dreg of the malt at the bottom of the tun; as soon as they begin to filtrate they flow transparent. Seventeen barrels are drawn, weighing eighty-six pounds of saccharine extract per barrel. The second and third mashes are laid on in succession, at the temperature of 185°,—the quantity of water required is nine barrels for each mash; they are stirred for ten minutes, and allowed to extract for one hour and a half for the second, and one hour for the third mash. The weight of saccharum obtained from the second is sixty pounds per barrel,—thirty pounds per barrel for the third mash,—consisting of eight barrels of wort for each.

Eight barrels of water, at 170°, are laid on for small beer, after the third mash. When the strong worts are in the boiler, they average sixty-six pounds per barrel, and are brought through to boil as quickly as possible.

The method of running the water through the malt, from the false bottom of the mash-tun, is very generally adopted in England. The best method of mashing that can possibly be followed, is to run the malt and water, simultaneously, into the tun, stirring and mixing as they descend. The ground malt is placed in a bin above it,

11

and a sluice regulates the quantity which is run down. By this method, the malt is completely mixed with the water, either by the machine or oars: there is also less danger of setting the goods, and the process is more effectually forwarded. This mode of mashing is well worthy the consideration of brewers who have not already adopted it.

The Edinburgh brewers have long since given up stirring by the machine, supposing that the oxide of iron tainted the worts during the process.

The machine is made use of in all parts of England. It is the most effective method, and no injury can possibly arise from its use if the size and weight of iron are any thing commensurate with the quantity of malt it is employed to work.

Where a machine is to be constructed for mashing, it ought to be made on the simplest principle, and with no more weight than necessary to do its work effectually. The motion of the blades or stirrers should be vertical; when horizontal, their action pushes the malt before them round the tun and injures the mash.

To small brewers the machine is highly important, as they have not a sufficient number of hands employed to make it convenient to work the mash with oars. The hand-machine is capable of mashing ten or twelve quarters of malt; and as the dimensions of a tun to mash such a quantity is that used by a numerous class of brewers, it deserves their attention.

A perpendicular shaft rises in the centre of the tun; on this shaft a tube is placed; a double wheel is fixed on the top, receiving motion from a bevelled pinion on the end of the horizontal axis of a fly-wheel at the side of the mash-tub, which is driven by hand, and which gives motion to the whole machine. Four arms project from the tube, on which are fixed the vertical stirrers; these are put in motion by two wheels on each side of the shaft acting on each other, the axles of which are the projecting arms from the movable tube, and which are sent round the mash-tun by the outer wheel on its top. The upper cogs of this are acted upon by the horizontal axis of the fly-wheel, which thus keeps the machine steady, and distributes the power which keeps it in motion.

The attention of the reader may be directed to several important points in the process of mashing.

1. *Sparging*, or sprinkling the mash, in such a manner as would give a stronger wort, without drawing it to so great a length as to render it necessary to boil down to strength for such a protracted time, which injures the extract, and drives off the essential aromatic bitter principle of the hops.

It requires very little explanation to render the mode of accomplishing this easily understood. Having laid on the first mash, and drawn it about one-fourth less than the usual quantity, the brewer lays on his second, with the same proportion of water. When he sets the tap of his second mash, and the worts begin to flow, he fixes the sparger at the same time, and sprinkles slowly as much water, at 180°, as of worts which flow from the tap. He can, by trying their strength, work out the saccharum as much as he may deem safe to preserve the quality of the ale.

This method would answer better in the English method

of brewing than sprinkling the first mash, especially when
the quick system of fermentation is followed.   The worts
obtained by stirring the mashes, and drawing them separately, are purer than when drawn by sparging the first
mash; although it must be granted that the slow fermentation causes the wort to cast a surer crop of yeast
than by the quick method, and much of the purity and
flavour of the ale depénds on the worts being judiciously
cleansed.

2. It is of very great importance that the division of
the worts into separate quantities, of different strengths,
so as to produce the different priced ales from the same
brewing, be well managed, when the establishment is of
much size, as ales of different prices are wanted to supply
the demand.   In such cases, the brewer must count a certain weight of saccharine extract per barrel, to represent
a certain price, and divide the worts obtained from each
mash accordingly.

By calculating how much wort strengthens in boiling
per hour, and also in cooling down for fermentation, and
how much of it is driven off by evaporation during these
processes, he is able to proportion the whole suitably to
produce given quantities of ale at the required strength
and price.

Of the best heats to be used in mashing,—of the quantity of water to be used for each mash,—and of the water
taken up by the malt,—these subjects have been noticed
previously; but it may be useful to recapitulate them, in
their proper place, while on the subject.

In the English system of ale-brewing, for mashing,
170° for the first, 185° to 190° for the second, and for

the third any heat from 170° to 180° will answer. It has been found that too high heats have a tendency to bring the ale sooner to acidity than when ranged low, although a stronger extract is obtained. The quantity of water to be used for the first mash, of course is regulated by the required strength of the ale.

It is estimated that malt takes up about 28 gallons of water for every quarter in operation. The brewer may, with this knowledge, lay on his first and second mash so as to obtain the quantity of wort required, according to the time he intends to boil. The Scottish brewers for the first mash never exceed two barrels of water to each quarter of malt, and sprinkle until they obtain the desired quantity for the brewing. The English brewers go the length of 2½ barrels to 3 barrels to each quarter, but which, as has been repeatedly mentioned, compels them to boil longer down to strength.

BOILING THE WORTS.—The strength of the worts are regulated by the process of boiling; and as it is during this process that the extract of the hops is obtained, to impart their aromatic flavour and bitter principle to the ale, it is always desirable in brewing to conduct it so as to produce the objects required—to arrive at the calculated strength as nearly as possible, and to preserve the *aroma* of the hops in the greatest perfection.

There are in the brewing on hand 33 barrels of worts in the copper, containing 66 pounds of saccharine extract per barrel, made from 80 bushels of malt; and there are 80 pounds of hops to be managed during the boiling process to extract the aroma and first bitter principle in the best manner, toward the preservation and flavour of the finished ale.

11*

The calculation is, that by boiling these worts for two hours, they will strengthen in the copper and on the coolers 20 to 21 pounds per barrel, and give 21½ or 22 barrels of ale, of the weight, before fermentation, of 87 pounds per barrel saccharine extract.

As has been shown, there are four actions that take place with malt in boiling, which materially affect it: 1. The process of boiling, which cuts the wort, as it is technically described, and coagulates the fecula of the malt. 2. Evaporation. 3. Concentration of the saccharum in proportion to the evaporation of the worts; and, 4. Destruction of part of the saccharum during the whole time of the boiling process. The reader's attention is again particularly directed to this, because it is upon such knowledge that the brewer is enabled to take advantage of the strength of his worts, and to vary his process so as to obtain the greatest possible quantity of finished ale, and to work to a required strength.

The proposition is: WORTS BOIL DOWN TO STRENGTH IN PROPORTION TO THE QUANTITY DRIVEN OFF BY EVAPORATION, AND TO THE WEIGHT OF SACCHARUM THEY CONTAIN; AND STRENGTHEN ON THE COOLERS IN THE SAME PROPORTION, DEDUCTING THE SACCHARUM DESTROYED DURING THE WHOLE OF THESE PROCESSES.

It was formerly stated that, during the process of boiling, worts strengthen one to ten per hour; that, when of the strength of 50 pounds saccharine extract per barrel, they increase 5 pounds in one hour; and when at 100 pounds, they increase 10 pounds. Therefore, in two hours' boiling, the first would increase 10½ pounds per barrel, and the second 21 pounds per barrel. The increase of strength on the coolers is in proportion to one-eighth part of the whole wort lost by evaporation, deducting the weight of saccharum which escapes during the process.

It is only by experiment that the exact quantity of wort driven off during one hour's boiling, and the weight of saccharum concentrated, can be determined. The subject is worthy the consideration of intelligent brewers, leading, as it does, to very important consequences.

The 33 barrels of wort, in the present case, are boiled for one hour, and then one-half of the hops are added, and the boiling continued one half-hour longer. The remainder of the hops is now delivered into the copper, and the whole boiled another half-hour, making two hours' boiling in all. If a lesser quantity of hops had been in operation, one-half for the whole two hours would have been boiled, and the other half for one hour; but, in this brewing, the present arrangement of their boiling is sufficient to extract the aromatic bitter principle in the finest state. Brewers who draw a large extract are obliged to boil down to strength; but they are not under any obligation to destroy the hop-aroma by boiling the hops all the time they boil the worts. Let any brewer try a simple experiment: let him infuse 1½ ounce of good hops in 1 gallon of strong ale wort, (which is about the proportion used in brewing,) brought up nearly to the boiling temperature, and keep it at that heat for an hour: he will find that the hops have imparted to the worts their fine aromatic bitter principle, without any boiling whatever.

It is well known that hops give up their essential aroma and first bitter very readily. One hour's boiling is quite sufficient. After being boiled for two hours together, most part of the essential oil or aroma and finest bitter are driven off, and the second or empyreumatic bitter succeeds, which is nauseous to the taste, and injures the ale.

When the worts have been finished in the boiler, it is always found advantageous to let them remain there for half an hour or so,—more especially, when the time of boiling the hops has been shortened. By giving the worts time, two benefits are obtained. The virtues of the hops to some extent are thus secured; and they lose a considerable portion of heat before being spread on the coolers;—thus saving a part which otherwise would be lost, by the rapid evaporation which takes place, when run down from the copper at the boiling temperature.

It is necessary to bear in mind that the form of construction of the boiler is an important consideration. The open boiler is still in use by many brewers. An upperback is of great utility; and, when properly constructed, becomes an object of value, as a larger quantity of worts, by one-third, can be boiled with safety when it is used. Boilers of the best construction are those used in London. They have a condensing-pan on the top, into which pipes are introduced, conducting the steam from the boilers beneath, which heat the water for the purposes of the brewery. Neither condensing-pan nor upperback are used by the Edinburgh brewers: the form of their boiler does not admit of it.

Brewers who require several boilers in operation, would do well to use the iron boilers. It may be important to determine the best form of construction of the iron boiler, which may soon come into general use. An iron boiler brings the worts sooner to the boiling temperature than a copper one, by one-fourth of time. It is easier kept clean; the oxide of iron is never present, the white fur with which it soon becomes incrusted, preserves it from the action of oxygen; while, in the copper boiler, its oxide is poisonous. It can be built at a comparative trifling expense, and is kept easily in repair. It is of importance to consider these advantages.

The form of the boiler is hemispherical, and its size is in proportion to the number of barrels of 36 gallons it is required to contain. It is made of ¼ inch boiler-plate, and the leggins which unite the crown are ⅜ inch. The boiler is enlarged with a leaden crib, which is fixed to its inner top by copper rivets, about an inch apart, all round; these, though small in the nail, have large flat heads to hold the lead close and prevent leakage. This leaden crib is of a convex form, being in diameter about a third more at its head than where fixed to the boiler. An upper-back, of a square form, is placed on the top, and which is capable of containing about one-eighth of the contents of the whole boiler. In fixing the upperback, a wooden seat is first placed round the head of the leaden crib. This is laid with whitelead and paper, on which the upper-back is placed; a strip of sheet-lead is run round the inside of the boiler-crib where the upperback and the wooden seat on which it is placed unite, and is nailed close with small nails, which keep the whole water-tight.[*]

The strength of the worts must be ascertained by the

---

[*] It is much the safest, for obvious reasons, to use some other substance than lead, and should always be done.

saccharometer before they are run from the boiler. Should any other method be adopted than that on the open coolers by evaporation, an allowance must be made to regulate the required strength, at the fermentation-point, for the saving effected. In proportion to the saving which arises from the worts not being exposed to evaporation, they must be made stronger in saccharum, and less in quantity.

COOLING.—Both in the English and Scotch systems of brewing, the method of cooling the worts by evaporation is about the same. They condense,—concentrate saccharum by evaporation,—lose bulk in proportion to such evaporation; and lose also a part of the saccharine extract, which escapes, more or less, the whole time of the process.

The worts are subjected to a loss of bulk by evaporation, from the time that they are brought through to boil in the copper, until they are cooled down to the temperature required for fermentation, and gain increase of weight in proportion, deducting such part of the saccharine extract as has been destroyed during the processes.

The weight of saccharum per barrel, contained in the wort, after the process of boiling has been finished, and before it has been run from the boiler, will be increased, when cooled down for fermentation, in proportion to what it acquires by evaporation on the coolers.

Therefore it is easy for a rule to be established for brewers to calculate the strength of their worts prospectively, from the time of obtaining them from the mash. The reader cannot fail to perceive of how much advantage this knowledge must be to those who require to make ales

of different strength, from the same malt in operation;—
or to vary the process of making a simple description
of ale, when the malt turns out a stronger wort than
usual; and thus secure an additional quantity of ale to
the brewer.

Whether or not a correct principle has been assumed in
dividing the processes of boiling and cooling the worts
each into separate actions, and attempting to prove the
loss of wort sustained, and the weight of saccharum acquired in each process, during the time of boiling and
cooling down for fermentation, must be left for competent
judges to determine; but the subject is so valuable to
brewers, that the investigation, perhaps, may arrest their
attention, and lead some of them to decide the question
by more accurate experiment.

Several improvements in cooling the worts were noticed
in describing the Scotch method of brewing ale; the most
important of which may again be spoken of.

1. *The Refrigeratory Cooler.*—A given quantity of
worts is run into a large tube immersed in water, which
cools them down to a required temperature, before being
spread on the adjoining coolers, to permit them to deposit
their sediment. These cooled worts are forced out of the
tube by another charge from the boiler, which is cooled
in succession; and in this manner the whole of its contents pass through the refrigeratory.

2. *Flat-surface Refrigerator.*—A flat, close receiver is
fixed round the inside of the coolers, and cold water is
kept constantly running into it. The bottom of this flat
receiver dips an inch into the surface of the worts, and

cools them rapidly down to the temperature required for fermentation.

3. The *Iron Coolers* and the *Refrigeratory Worm* will be found of much utility during the summer months, and are worthy the consideration of brewers who have not yet adopted these methods of cooling.

During the process of brewing, as generally conducted, when the whole quantity of worts obtained from the different mashes, and the weight of saccharine extract they contain are ascertained, and afterward compared with the quantity and weight brought down to fermentation, the loss of both worts and extract are correctly proved.

It is an object of very great importance to economize the processes of boiling and cooling, without injuring the quality of the production, the whole being resolved into this:—*From a given quantity of malt and hops, to bring the greatest weight of saccharine extract in solution to the temperature required for fermentation with the least possible loss.*

It must be readily seen by every one, that this is the only rational view that can possibly be taken of the subject.

FERMENTATION.—There is no part of the process in the English and Scotch methods of brewing which differs so much as in that of fermenting the worts. The difference has before been shown to arise from pitching them to ferment at different temperatures, and with different quantities of yeast. By the English method, a temperature of 60° to 65° is chosen. The English brewers are extremely cautious with their heats of fermentation; and the intelligent have generally adopted the mean of these heats,

62½°, or for practice 62°, as the best that can be adopted during all seasons of the year. By the Scottish method, as formerly mentioned, the heats are from 50° to 55°; and 52½° is the mean, or 52° in practice, 10° being the difference of quick and slow fermentation.

In both methods, the chemical principle is the same; the difference of heats used causing a different action of the worts in their time of resolving into alcohol and throwing off the yeast and impurities they contain.

Observe one important fact. The fermentations at the mean of these heats, 52½° and 62½°, *are equivalent*, producing, by different modes of action, the same results.

When worts are fermented at 52½°, in eight or ten days a liquor is produced, combining alcohol and starch-sugar in such proportions as to constitute strong ale of a certain flavour and quality; and worts fermented at 62½° produce a liquor of somewhat different flavour, but combining alcohol and starch-sugar in like proportions, all things being equal in the quality of the malt and strength of the worts: in both instances, the heat of the fermented worts rising to nearly similar temperatures,—10° or 12° above those at which they were set to ferment.

If the worts were pitched to ferment at 72½°, in place of 62½°, they would spring rapidly to a high temperature before a sufficient quantity of alcohol was evolved. No doubt they would bear yeast, but it would be deficient in those requisites which distinguish strong healthy yeast; and when tunned and cleansed, the ale would have a thin, vapid taste. It is a subject of too much importance to be hastily passed over. It is either true in principle or it is not. It is quite evident that, in the quick and

12

slow methods of fermentation, there must be equivalent heats. All that can be done, however, while the question remains unsettled, is to endeavour to fix on the best heats to produce the best results. The worts in the coolers being all in readiness, nine gallons of rich strong yeast having been thrown into the gyle, about a barrel of worts is pitched and thoroughly mixed with the yeast. The worts are now run into the gyle in full flow, until the whole have been obtained.

Within twelve hours, a head will have gathered on the worts, the healthy state of which cannot be mistaken. In a few hours a faint white line breaks round the head of the gyle; afterward the surface begins to show white spots, which soon increase and rise on the surface.

The head is formed in from twenty to twenty-four hours; it consists of white froth, at first shooting up in little pyramids, and then uniting and swelling up with a very slight and scarcely audible singing noise, which is a favourable sign that all is well.

This white froth, in the course of a few hours, is changing into gas bubbles, caused by the carbonic acid escaping from the worts: these soon appear numerous, and enlarge into each other. The head assumes a cream-coloured tinge, and the formation of viscid yeast has evidently commenced. The fermentation proceeds; the head of the worts still bearing up the yeast, which swells up toward the centre. The heat of the worts is 70°. The flavour from the gyle is cool and sweet, and this is more perceptible when fine new malt is in operation. The appearance of the head of the gyle now assumes the look of the worts coming to maturity. It is decisive in about thirty-six hours.   The yeast is of a rich, close texture, but without rolling over in a broad, unbroken mass, as in the Scottish gyle during slow fermentation.

The head of the gyle now begins to sink, and the yeast to slacken round the edge; but the brewer need not hurry the process; two or three hours at this time are of essential service to strengthen the yeast in cleansing, but it must not be permitted to sink much.   Four pounds of flour and two pounds of salt for every twenty-five barrels is a good proportion of these substances, to be mixed with the worts, before tunning, to season the ale and increase the fermentation in the barrels.   The heat of the worts is now 72°, and the gyle is thought to be ready.   The flour and salt are thrown in by handfuls over the yeast on the gyle, and the contents are plunged and mixed together.

When this is accomplished, which occupies a minute or two, the whole is run into barrels placed in the tunning-room, to undergo the process of cleansing.

During the quick process of fermentation, it may be observed that the head of yeast in the gyle is never touched until beat in at last, when the flour and salt are mixed with the worts.   Were it beat in as in the Scottish process, the worts would spring into so rapid a fermentation, from the quantity of yeast with which they were pitched, and from their high degree of temperature, as to endanger the brewing.

CLEANSING.—The continuation of the process of fermentation, by mixing the yeast and worts together, and overturning the whole contents of the gyle into casks, from which the worts partially escape in froth with the carbonic acid, and are again filled at intervals of two hours for twelve hours, and then every four hours, until, by repeated working out and refilling, the yeast comes to maturity and separates from the worts—this is called *cleansing*.

It has been already observed that, in the Scottish method, this term signifies, merely, the running the finished ale into casks, where no more fermentation takes place; but the present method is a continuation of the process, which requires much care and judgment in the brewer to bring to a successful termination.

Before being placed on the stillions or troughs, the casks should be examined singly, to ascertain that the bung-stave is good,—the spile-holes filled,—the tap-ends all right; and that the casks are sweet, and in good condition to serve their purposes.

Every thing being right in the tunning-room or cellar, the first operation proceeds, or removing the worts from the gyle into the casks; for doing which there are several methods: 1st. By carrying the worts from the gyle in a pail, and filling the barrels progressively, sufficient worts being left to finish the cleansing. 2d. By means of a hose made of leather, long enough to reach to any part of the cellar where the barrels are placed. One end of the hose is open, to admit the tap of the gyle. A stop-cock is fixed on the other, with a small conduit to enter the bunghole, and to prevent the spilling of the worts. 3d. By the filler-trough. This is a small trough laid across the tops of the barrels, the whole length of the stillion; into each bunghole a filler descends from the trough, and a stopper is fixed by a small chain to each division of it, where the fillers are placed to be instantly

ready when wanted to stop the flow of worts when the
barrels are full.   The worts are conveyed to the trough.
by the hose already described, and which is still kept in
use.   4th. By the cleansing-vat, with float-valve.   This
method of cleansing is confined to the London porter-
breweries.   The fermented worts are distributed into the
barrels by means of the cleansing-batch being placed on
a proper level with the casks, which are kept constantly
full, on the same principle that regulates the common
water-cistern, fitted with a ball-cock or floating-valve.
The casks having been all carefully examined, and placed
on the stillions, the filler-trough is laid over the first
division of barrels, and, every thing being right, the tunning proceeds.   The filler-trough is, to a certainty, the
best method of cleansing in practice; and brewers who,
are judicious enough to appreciate the value of quick,
clean, and economical work, would do well to adopt it.

A gauge of the gyle is taken, when we wish to tell the
number of barrels to be tunned.   The brewer then places
such a number of casks on the stillions as the quantity of
ale in the gyle requires.

After the casks are filled, which, when the filler-trough
is used, is very expeditiously done, the overplus of the
wort in the gyle is run into the *plustub*, as near as can be
guessed, to the quantity of a hogshead of 54 gallons for
every twenty barrels of ale in operation.   With the liquor
on the stillions, this is sufficient to keep the barrels in
fermentation until cleansed.   The wort flows out of each
barrel in a stream of froth, composed of worts under the
action of carbonic acid.   This stream accumulates in the
stillions beneath, and, the carbonic acid escaping, the

12°

froth is again resolved into wort, which is taken off the stillions by a spigot or stopcock, and filled into the barrels every two hours with wort from the plus-tub, until the yeast separates from the worts, and the fermentation has ceased altogether.

In finishing the ale on the stillions, the brewer must not trouble himself with filling up with mild ale of another brewing. It does no good whatever. If justice has been done to the operation in hand, it is independent, and requires no assistance.   Still, it is very useful to reserve a firkin or half a barrel of wort of the same brewing in hand; and in place of filling up with refuse of the plus-tub and stillions, and doing damage, the brewer has thus fine worts to finish with and preserve the purity and flavour of the ale.

BREWING OF PORTER.

UNDER this head will be given a short statement of the method of brewing porter, which will be practically useful, premising that no deleterious ingredients are used; for every intelligent brewer will acknowledge that, whether in the manufacture of porter or ale, the best malt and hops are the cheapest materials that can be employed.

It may be stated, however, that there are various substances, which are used, such as roasted malt, burned sugar, malt-wort burned down, (*essentia bine,*) liquorice, orange-powder, heading, and finings; but all these additions are simply to improve the flavour, or to preserve the keeping quality of the liquor, and not as substitutes for malt.

As porter is made by some brewers in the same utensils as those in which they make their ales, it is unnecessary either to repeat a description of them or to notice their dimensions; but it may not be out of place to state to the reader the method of making porter-malt; as every brewer should have some knowledge of the mode of preparing it, —although it has been nearly displaced, in large establishments, by the more profitable use of pale malt in porter-brewing.

There is quite a difference in the genuine porter-malt kiln and the common description for drying pale malt. The floor of the former is laid with tiles in the usual manner where such are used. The fire-chamber beneath is built of brick, within the square apartment, in the shape of an inverted pyramid, in the apex of which the furnace is placed. The furnace is arched with fire-brick, and extends 2½ feet within the chamber, to disperse the heat equally to the floor above.

The malt to be prepared for porter-brewing is half-made in the usual manner for drying pale malt. It is then divided into two or three parts, which are dried and finished on the kiln at such a high temperature as speedily turns it of a brown colour, but without scorching or charring it; and converts it into porter-malt.

It is first dried in the usual manner. Birch-cuttings, or beech when the former cannot be procured, are prepared to blow it, as it is termed, on the kiln, and give it the brown colour and that bitter principle which is so desirable to the taste in the consumption of porter.

When the malt is spread on the kiln-floor, the furnace is gradually charged with the wood-cuttings until a temperature upward of 200° is obtained.   The maltster carefully watches it until it begins to burst by the escape of the air confined between the kernel and husk of the grain. He now turns it, and, with his assistants, by means of shovels and brooms works it quickly, and sweeping each division, as it is proceeded with; and this process is repeated until it is judged sufficiently brown for its purpose.

Its germinating principle is destroyed by this incipient charring, and it loses the capacity of yielding sugar, by mashing, in the proportion of twenty per cent. to pale malt made from the same description of barley.

MALT AND HOPS.—The malt required to make 20 barrels of stout porter is 40 bushels of pale and 20 bushels of genuine porter-malt.

Each of them are rather small ground and mashed in the same tun.   They are mashed separately in large breweries; but on a small scale it is not necessary.   The amount of hops employed is seventy pounds.   The other ingredients used will be described during the process.

MASHING.—As stated previously, 40 bushels of pale and 20 bushels of genuine brown malt are employed. The quantity of water for the first mash is 20 barrels, heated to the temperature of 180°, which is stirred for twenty minutes, and stands three hours to extract, the head of the tun being covered.

The tap is set and the worts flow into the underback,

after three hours. The heats of the mash are about the same as in ale-brewing, 140° at the surface, and 148° to 150° at the tap. Fourteen barrels of wort are drawn, weighing 66 pounds extract per barrel. Fourteen barrels of water at 190° are laid on for the second mash, which is stirred for ten minutes, and allowed to extract for two hours. From this mash 12 barrels of wort are drawn at 84 pounds per barrel. The first mash having previously been pumped into the boiler, 8 barrels are laid on for the third mash at 180°, which, after standing to extract one hour, are recovered at 20 pounds per barrel.

BOILING.—The whole quantity of worts now in the boiler are 34 barrels, weighing 44 pounds of saccharine extract per barrel, and which are brought through to boil. as speedily as possible. In porter-brewing, sprinkling the mash does not succeed; and, in mashing, the temperature of the atmosphere is a matter of indifference, care being taken that too high degrees of heat are not used, although brown malt bears a higher temperature than pale. With brown malt, the heat of the water should never exceed 190°. Eighty pounds of hops are added, or delivered into the boiler, when the worts have come to the boiling point.

The time of boiling the worts is regulated by the quantity drawn from the mash. In the present instance, two hours and a half will be required. The ingredients to be mixed during the process are Spanish juice, burned sugar, clarified sugar, and liquorice root. 1st. Burned sugar. Thirty pounds of good raw sugar are put into an iron boiler which has a circular bottom, and dissolved in one gallon of boiling water over a moderate fire. It must be kept stirred, and attended to with care. When it has boiled a few minutes, and been stirred with an iron scraper, it thickens, and acquires a bitter taste. Care must be taken that it does not get scorched. It must be constantly stirred, and a small quantity of hot water added, to keep it from setting on the bottom of the boiler. It is ready to be removed when it approaches to inflame, which is done by thinning it with boiling water, and delivering it into the boiling worts. 2d. Spanish or Leghorn juice. Six pounds weight are broken into small pieces, put into a net, and sunk into the worts in the boiler, to be dissolved during the process of boiling. The net is hung from the top, and must be put in when the worts come ·through to boil; or the liquorice may be dissolved previously, and thrown into the copper as soon as the worts come to the boiling temperature. 3d. Twenty pounds weight of clarified sugar, broken into small pieces, are mixed with the worts. And, *lastly*, three pounds ground liquorice-root are added.

When these materials are put into the copper, the worts are boiled rapidly for two hours and a half, when the quantity and weight must be ascertained, and a judgment formed of the time for their being run through the hopback into the coolers.

The third wort is often boiled separately without hops, and run into the hopback after the worts from the main copper have been strained; by which means the strong kind taken up by the hops is transfused by those from the third mash. From what has been already said, it will be evident that this method deprives the worts from the third mash of their share of ingredients in the boiler, although, from the quantity of hops used, it becomes an object to transfuse them.

The question for the brewer to decide, is whether the transfusion of the hops will compensate for the deterioration of the worts from the third mash not having been boiled together with the first two worts, to obtain their share of the foreign ingredients necessary to flavour and enrich the porter.

COOLING.—The process of boiling the worts, and the laws which regulate that process, having been explained in the description of ale-brewing, it is unnecessary again to enter on the subject.

The porter worts being spread on the coolers, are cooled down by evaporation to 62°, being the same temperature for the fermentation of ale-wort by the quick method, by which porter is always made.

The laws which govern the brewing of all worts also regulate those of porter. In the boiler and on the coolers they strengthen in proportion to the water driven off by evaporation; and the prospective increase of and loss in quantity may be calculated when the worts are drawn from the mash, on the same principle as that which guides the brewer in the production of ale.

FERMENTATION.—Twenty-two barrels of worts, (when cooled down to 62°,) containing sixty-five pounds saccharum per barrel, are pitched to ferment with nine gallons of yeast. The best yeast for onset is fresh strong-ale yeast. To cause it to strike well just before being thrown into the gyle, an ounce of the best ground ginger is mixed with it; but when the brewer is sure of the goodness of his onset, this is unnecessary. The fermentation of porter worts runs the same course, and is similar in chemical action to that of ale; but when the head of yeast is formed, and comes to maturity on the worts, it does not support itself there so long as ale yeast; it has not the same capacity of holding sugar in partial decomposition as the latter, and were it not beat in after full formation, it would sink down through the worts and leave the sur-face clear; but in this latter case, which would ruin a brewing of ale by yeast biting, and subsequent vinegar formation, the porter, although deprived of the cleansing process of fermentation, is not lost. It is preserved in a vinous state, by the strong combination of hop-bitter and burnt sugar. It is stored in a vat with finished porter; and, without deteriorating the quality of the latter, re-covers its own value.

When porter is made with fresh malt and good hops, and when the ingredients have been judiciously prepared, the worts under process of fermentation possess a full taste, rather bitter, but grateful to the palate, and the gyles send forth the flavour by which porter, as a malt-liquor, is distinguished.

After the worts have come to maturity, which is known by the head of yeast beginning to slacken round the sides of the gyle, and to drop a little in the centre, four pounds of flour and two pounds of salt are thrown by the hand over the yeast, which causes it immediately to shrink to-gether. The worts and yeast are then plunged and mixed thoroughly, and the process of tunning commences.

CLEANSING.—Pipes or butts are always preferable to barrels in the cleansing of porter. The fermentation goes on better in large casks. It is not so strong as that which

takes place in cleansing ale, and requires to be more frequently filled up. In cleansing porter, the casks must be
filled up every half-hour for twelve hours—then every
hour for the same length of time—then every two hours;
and until the fermentation ceases altogether, every four
hours. The stillions must not be drawn too close, as the
yeast would clog up the run of the liquor, and loss ensue,
care being taken that there is sufficient worts in the *plus-
tub* to finish the brewing.

STORING.—When the cleansing is completed, the future
disposal of the porter determines the brewer what course
to pursue,—either to remove the brewing into cellar-
stock, in the same casks into which it was tunned, to be
afterward racked into barrels, to be sent out to customers,
or to store it in a vat for the same purpose.

In the first instance, supposing the porter to have been
made in the spring, to be ready for delivery during the
summer, such old ale in the brewery as may be fit for the
purpose is put down with hops,—two pounds to each barrel,—and kept until required. Nine gallons of the old
ale are first delivered into each barrel of thirty-six gallons,
and smaller casks in proportion, and the barrels filled up
with porter, when racking it from the pipes into barrels.
A table-spoonful of heading is added: this consists of
equal parts of alum and carbonate of potash; also the
same quantity of orange-powder; last, about the third
of a pint of finings made up with vinegar and isinglass.
It is customary with some brewers to keep a supply of
London porter by them, and add a gallon or two to each
barrel racked; and it may be easily imagined, that other
productions will be all the better for such an addition.

One thing is absolutely necessary in brewing porter : the brewer should contrive to make a strong-bodied, fine-flavoured malt-liquor, which, if it does not pass for *" London porter,"* will pass on its *" own merits,"* which is all that is required.

In the storing of porter, it is necessary that a quantity equal to three-fourths of the measure of the vat should be overturned into it fresh from the stillions, immediately after having undergone the process of cleansing. One-fourth of prepared old ale is added to fill up. The man-hole is battered down air-tight, sufficient space being left for the rise and fall of the liquor within,—a slack-pile or vent-plug being inserted at the same time on the head of the vat.

A trial stopcock is placed about a third part up, by which the brewer has access to judge of the progress the contents are making toward ripeness. It is ready in two or three months, when the same additions are made in racking as previously described in racking from the pipes.

At this stage, brewers greatly differ as to the additions to be used to obtain the flavour of " London porter." Decoction of bark and porter extract are sometimes used ; but, as has been already mentioned, if the malt-liquor is really good, there is not much occasion for using such ingredients.

The preceding description of brewing malt-liquor must be taken, not as the general method adopted by all brewers of making porter, but rather as affording general information of the mode of conducting the process, and preparing the foreign ingredients, by a judicious management of which an imitation of " London porter " may be produced.

If *essentia bine* is preferred by the brewer to burned sugar, for colouring, it is made by boiling down fifteen gallons of strong-ale wort, of the strength of seventy-two pounds saccharine extract per barrel, to the consistence of molasses, and then carefully treating it over a slow fire, until it thickens and acquires that bitter taste required for the purpose intended. Fifteen gallons of worts boiled down is sufficient to make twenty to twenty-five barrels of porter.

If sugar should be subjected to a slow heat and dissolved, it crystallizes on cooling, and loses the capacity of granulation; but on being dissolved in water, it is still subject to fermentation, although not with such rapidity, or to such a degree, as common sugar or glucosin, (malt sugar,) hence the motive for using it in porter-brewing. But when sugar is boiled down to make colouring, and subjected to the continued action of caloric, it loses both the capacity of crystallization and fermentation, and becomes a vegetable bitter. When it is reduced to this state, it is dissolved in the porter worts, giving them a dark brown colour; and, not being subject to the fermenting principle, remains dissolved in the finished liquor, preserving its transparency and bitter quality. Essentia bine, or malt-wort, boiled down and burned, possesses the same qualities, and gives to porter a similar property as sugar colouring. Before the Spanish-juice is mixed with the malt-wort in the boiler, it should be dissolved in water. When dissolved in the worts during the process of boiling, it causes the liquor, afterward, to become ropy, more especially when put down with old ale. The peculiar flavour of liquorice or liquorice-root is slightly perceptible in every species of porter properly made.

A description of making up a vat of sixty barrels of porter will here be given, which will be found quite full for every purpose.

Two brewings of porter, consisting each of twenty barrels, being required,—sixty bushels of pale and brown malt in equal proportions, with fifty pounds weight of hops, thirty pounds of essentia bine, and the same quantity of clarified sugar, with six pounds of Spanish juice and four pounds of ground liquorice-root, is used for each brewing. They are conducted by stirring each mash; and quick fermentation; and cleansed in wine-pipes.

Two days after being cleansed, the whole is overturned into a vat, and fifteen barrels of fine old ale, previously prepared with two pounds of hops to each barrel, are added. The vat is headed with five barrels double-stout London porter, room being left for expansion; the man-hole is battened down, and a vent-tube properly inserted on the head of the vat. It thus contains sixty barrels of porter, averaging, with the double-stout, sixty-six pounds saccharum per barrel. The porter remains to ripen for three months,—from April to July,—when it is racked off, to order, and turned out in remarkably good condition. Nothing is added but the usual heading of carbonate and orange-powder, and about one-third pint of finings per barrel when sent out to customers.

A description of the best method of managing returned ale should be spoken of before closing the remarks on porter-brewing on a small scale,—either for the purpose of porter-brewing, or to preserve it, to be resold as old ale, which often becomes acceptable during the summer months. There are few brewers but who have ale, on the second fermentation, occasionally returned by their customers, which is often a cause of reflection, as useless as the loss is serious, if not immediately remedied. Under such circumstances, the best thing a brewer can do is to remove and replace it with stock in better condition.

If there be a considerable quantity on hand of the ale returned, part is selected to be used in porter-brewing; or the whole is made up for keeping, or old ale. The first is mixed with two pounds of hops to each barrel, and placed aside until required. The kind for keeping is overturned into wine-pipes, which, when obtained fresh from the wine-merchant, do not require to be washed, but the ale may be overturned into them on the wine-lees. The pipes require a quarter-hoop and chime on each end, to render them secure and serviceable for brewing purposes.

When it is desired to overturn the ale into the pipes, a barrel of the strongest kind, new from the tunning-room, is added to each pipe, with 4 pounds of best hops. A slack-bung is put in each, and they must be kept full and carefully attended to.

Before long this ale acquires the flavour of home-brewed, and is sold during the summer months, in fine condition. Some brewers use carbonate of potash to take up the acidity of old ales; others put a barrel or two of old ale into the gyle before tunning. Injury is done by mixing them in both casks, which is more hurtful to the

brewer, in the end, than any immediate advantage he
may derive. In porter-brewing, none should be used except what has been carefully prepared and is quite fit for
the purpose.

———

ALE BREWING ON A SMALL SCALE, SUITABLE FOR BREWERS IN THE INTERIOR OF THE STATES.

In a brewing of two barrels, or 72 gallons of ale, 9
strikes (bushels) of malt, with 12 pounds of hops, are
used; but the quantity of the latter may be two pounds
more or less, according to taste. The malt is ground small
and mashed with 72 gallons of water at the temperature
of 160°. The malt and water are stirred completely, and
allowed three hours to extract, the mash-tun being closely
covered up with sacking. Forty gallons of wort are drawn,
into which the 12 pounds of hops are mixed and left to infuse; meantime, 60 gallons of water at 170° are run into
the tun for a second mash, which are drawn off after standing two hours. The worts are boiled together for two hours,
and, after being cooled down to 65°, strained through
a flannel bag into the fermenting-tub. One and a quarter
gallons of yeast is then mixed with the worts, and left to
work from 24 to 30 hours. When run into the barrels to
cleanse, 6 or 7 gallons are reserved for filling. About
18 gallons of beer are made after the second mash is

drawn, and the same hops used as were boiled with the ale-worts, or fresh hops, if such are preferred.

The quantity of malt and hops used in brewing strong ales must be in proportion to the quality and strength of the kind required. The general calculation, when speaking of the strength of ale in England, is to name it at a given number of: bushels of malt per hogshead, which contains 54 gallons, or 1½ barrel.

The range of measure of malt for this kind of ale may be taken from 6 to 10 bushels per hogshead; above the latter quantity, it is mere waste to attempt to mash, except it is intended to keep the ale a number of years. But in further explanation of this, the reader is referred to *Fermentation* in the Scottish system of ale-brewing.

ALE MADE BY PRIVATE FAMILIES IN SOME PARTS OF ENGLAND.

This brewing is done on a small, economical scale, and is well adapted to many private families and others in this country. There cannot be the slightest doubt that ale of a fine quality can be made at quite a moderate expense, thus rendering it an object of domestic economy.

It does not cost much to obtain all the utensils and other requisites for brewing on a small scale. A wine-pipe or rum-puncheon, sawn in two, makes an excellent mash-tub and gyle. A boiler made of iron-plate, capable of holding 36 or 54 gallons, and a cooler in proportion, can be procured at a very reasonable price, in every town or village of any size; these, with two or three casks, and other requisites procured, necessary for the purposes of brewing, the whole may be completed and fitted up in any small out-house that can be conveniently adapted as the brewhouse.

Three bushels of malt and 2¼ or 3 pounds of hops are sufficient to make 36 gallons of ale of good quality. By dividing these materials, 18 gallons may be produced of equal strength. A few gallons of small beer are obtained in both cases, which, with the grains and yeast left after each brewing, every family, especially those in country residences, can find use for.

It will be observed that three bushels of malt produce 36 gallons of ale equal to 72 pounds of saccharine extract, which is a strength of good ale for family use. One pound of hops to each bushel of malt in operation is sufficient in all cases.

In the description of brewing on a small scale, for family use, it is unnecessary for the reader to be troubled about the density or specific gravities of the worts, or weighing them by the saccharometer, or with technical or chemical terms. A thermometer, however, to ascertain the heats, is a very safe guide to make sure work with, and ought to be found in every brewhouse.

In a few plain directions, all the information on the subject necessary can be conveyed, by which the process may be carried through successfully, and ale of the best quality produced.

The first essential in brewing is a minute and careful

inspection of all the utensils, casks, and other requisites; every thing must be previously arranged and in proper place; and each and all of them scrubbed and scalded with boiling water repeatedly, until the whole apparatus is per-- fectly sweet and clean.

Let us, therefore, suppose that the whole utensils are perfect and in readiness, and that 36 gallons of ale and 10 gallons of table beer are required from three bushels of malt and three pounds of hops. One hundred gallons of pure soft spring-water must be provided; and the boiler filled and brought to the boiling heat. Waiting until the water cools down to the heat required for mashing, is better than tempering it with cold water. When at the temperature of 180°, 32 gallons are run into each mash- tun. The water loses above 5° in going down; if the malt be small ground, it is better to let the water run into the mash-tun be 170° to 175°. The three bushels of malt are now added, one bushel at a time, and immediately stirred and mixed thoroughly with the water; the whole operation taking a quarter of an hour or twenty minutes. The mash-tun is then covered, and allowed three hours to extract. In the mean time, the remainder of the water is delivered into the boiler, and prepared for the second mash. After three hours the worts are drawn from the mash, slowly at first, until they run transparent. About 22 gallons of worts are obtained, the remainder being taken up by the malt and left in the mash-tun. The se. cond mash is now commenced by letting down 34 gallons of water at 180°, and immediately stirring it for ten minutes.

As in the former case, the mash is then covered up,

and allowed two hours to extract.   The water in the
boiler is now prepared for the small beer, and when it
comes to the boiling heat, it is run into a clean receiver,
—and the worts of the first mash delivered into the boiler.
The second mash is then drawn, and also removed into
the boiler, which now contains 54 gallons of worts, and
must be brought through to boil as quickly as possible.

The worts for the small beer are now laid on to the extent of 14 gallons, at any heat from 160° to 170°, and
allowed one hour to extract.   The 54 gallons of worts
in the boiler are boiled for half an hour without the hops,
they are then added, and boiled another hour, which is
quite sufficient to extract their aroma and first bitter principle.   The fire is then drawn, and the worts allowed to
remain in the boiler twenty minutes, or half an hour, to
infuse the hops a little more, and prevent too rapid evaporation on the coolers, were the worts spread on them
at a boiling heat.

The worts are then strained through the hop-drainer
on to the coolers, and cooled down to 65°. Sixty-two degrees is a better fermenting-heat, were a larger brewing
in hand; but on a small scale, 65° is judged necessary, in
consequence of the heat lost by using small utensils. For
36 gallons of worts, half a gallon of yeast is requisite. It
must be thick, strong, and fresh, and obtained from a
brewing of equally strong worts, or stronger than those to
which it is to be applied. The worts are first run into
the fermenting-vessel, and the yeast then put in and
mixed together.

In thirty hours, or thirty-six at furthest, the fermenta-

tion will have come progressively forward, as described
formerly; and when the head of yeast begins to sink, the
worts are tunned into casks, as suitable for family conve-
venience.

Three or four gallons of the fermented worts are reserved to fill up with. The casks work out into a receiver
or trough, and are kept filled up every hour and two hours
until the yeast separates and the fermentation ceases.
The 54 gallons of worts taken from the two mashes are
reduced to 36 gallons of finished ale, by waste arising
from evaporation in boiling and cooling down for fermentation, and in cleansing.

Refined sugar is sometimes added to help the worts;
but when there is plenty of malt, there is no occasion for
it; except, indeed, where too much wort is drawn from
the mash, and threatens weakness, 6 pounds of refined
sugar will greatly improve it, and may be put into the
boiler with the hops.

The beer-wort is boiled for an hour after the strong
worts, and the hops used in the brewing are boiled half an
hour, or a quarter of an hour will do. When ale is made
from 4, 5, or 6 bushels of malt to a barrel of 36 gallons,
a strong table beer or weak ale may be made, in which
case it is requisite to use 1 pound of fresh hops, and boil
them an hour with the wort. One and a half pound
of fine hops are put into each barrel of 36 gallons, when
ale is required to be kept over the summer—the bungs
of the casks slackened, and the casks kept full.

Where a smaller quantity of ale is required to be made
than 36 gallons, the materials are divided in proportion.

On whatever scale the brewing is conducted, the best malt and hops are requisite to produce malt liquors of the best quality; and it is almost an infallible proof of skill when, to the eye of a judge, every thing about the brew-house appears to be arranged and conducted with cleanliness and economy.

ALE MADE BY FAMILIES OF DISTINCTION IN SOME PARTS OF ENGLAND.

THIS constitutes, I suppose, the finest ale that is made, but is attended with too much expense ever to become extensively manufactured in this country.

In brewing ale of the best description in any considerable quantity, it may at once be supposed that a commodious brewhouse exists, fitted up with every requisite for its purpose, and that a brewing is to be made of nine hogsheads of ale, from nine bushels of malt to each hogshead, with one pound of hops to each bushel of malt.

In laying on the first mash, it may be assumed that eighty-one bushels of malt take up five hogsheads of water to saturate the malt. The calculation to be made is, how much water must be laid on, in addition, to extract the saccharine matter or worts necessary to make up the quantity required to produce nine hogsheads of finished ale; making an allowance for the waste that takes place, from the water being taken up by the malt and by evaporation, in boiling and cooling down for fermentation?

Eighteen hogsheads of water will be required for the brewing; but it is certain that, were the whole laid on at once, a large proportion would be left in the mash, and the malt prevented from yielding a further extract. Separate mashing is therefore necessary, to obtain sufficient quantity of worts, of strength commensurate with the weight of malt in operation.

Twelve hogsheads of water, at a temperature of 170°, will be required for the first mash, which, after being stirred a quarter of an hour, or until the malt and water are thoroughly mixed, is allowed three hours to extract. Seven hogsheads of worts will be obtained, weighing one hundred and thirty pounds saccharum per barrel of thirty-six gallons. Six hogsheads of water, at 180°, are laid on for the second mash; which, after being stirred, and allowed two hours to extract, yield six hogsheads of worts, weighing eighty pounds per barrel. Five hogsheads of water, at 170°, are now laid on for table-beer, which, after infusing one hour, are drawn at the weight of thirty pounds per barrel.

BOILING.—Thirteen hogsheads of worts are now in the copper, weighing one hundred and three pounds saccharum per barrel, and brought to the boiling temperature as rapidly as possible. Care must be taken to obtain the hop extract of the finest flavour. After the worts have been boiled for an hour, forty pounds of the hops are delivered into them, and boiled another half-hour. The remainder is then added, and boiled rapidly half an hour more; making two hours boiling for the worts, one hour's boiling for half of the hops, and half an hour for the remainder. This is quite sufficient to extract the aroma and first bitter principle, and impart to the ale that fine aromatic flavour which it should possess when finished in the highest state of perfection.

After the worts have been boiled from one and a half to two hours, according to the quantity to be boiled down to strength, the fire is drawn, and they are allowed to remain in the boiler for half an hour. This arrangement allows the hops to infuse the better, and reduces, at the same time, the capacity of the worts to escape by evaporation, which is greatest at the boiling-point.

Before being pitched to ferment, the worts are drained through the hopback, and cooled down in the usual manner. This heat is 2° below the standard; but with these strong worts it is prudent to be under the common brewers' heats, whose ales are weaker in saccharine extract. The quantity of ale required is nine hogsheads, or thirteen and a half barrels, and the weight in saccharine extract is one hundred and thirty-four pounds per barrel.

Six gallons of yeast is the quantity for ferment. Half a barrel of worts is run into the gyle-tun, with which the yeast is mixed; and the remainder of the worts immediately pitched in full flow from the coolers at 60°, as previously mentioned.

With the exception that during the carbonic acid formation the froth rises on the surface of the holder, and sometimes swells up to the head of the tun, the appearance of the gyle becomes similar to that of common ale. This is always the sure sign of a healthy, vigorous fermentation. From having been pitched at 60°, the worts take more time to come to maturity for cleansing; but they are generally ready within forty-eight hours. One pound of salt and three pounds of flour are thrown into the gyle before cleansing, and, as usual in quick fermentation, the yeast and worts are plunged and mixed together.

The worts are cleansed in pipes set on their bilge on stillions. The common method of filling up when working out until they come to yeast, is generally followed in making home-brewed ale. The pipes must be filled up every hour for the first six hours; then every two hours, until the worts come to form yeast,—then every four hours until cleansed.

The ale arrives at the stage of what is called the "first fining" within six or seven weeks after cleansing; and may be bunged down and brought into use. But it is a general practice, in family establishments of magnitude, to hop down the new ale for keeping; the ale used from the cellars being in succession of stock, generally from six to twelve months old. Six pounds weight of hops are put into each pipe of new ale, which is reckoned sufficient to keep it over the summer season. The bungs are always out for the first two or three months. The cask being kept full, and the hops remaining at the surface, render driving the bung unnecessary, and prevent the ale running into the acetous state.

In all cellars where a large stock of ale is kept, the bungs of the pipes must be raised before the heats of summer, and one and a half pounds of the best hops put into each, as required; they are kept slack-bunged, with a vent-plug in each cask. Ale, in bottle, should never be kept longer than the season in which it was intended to be used. But when in wood, it may be preserved in fine condition for many years. The judicious addition of hops, seasonably made, is the means for its preservation. When ale has been more than one year in the cask, it may be renewed as follows:—The contents of two pipes are overturned into three empty ones, which latter will require about thirty-six gallons of new ale, each, to fill them up; care being taken that the old ale is racked off pure and free from the hops which had been put into it in former years for its preservation. The new ale must be as strong, or stronger if possible, than that to which it is added; and it must be added in a day or two after being cleansed, and before it is removed from the stillions. Three pounds of the finest hops are to be put into each pipe, which are bunged down, and placed on end in the cellars. This ale, supposing that both the old and new are made from nine or ten bushels of malt to each hogshead, when ripe, approaches to wine in quality, and affords, in all probability, the best sample of " HOME-BREWED ALE."

SMALL BEERS.

THIS is one thing concerning which but few writers on brewing have said any thing definite or instructive. Whether it has been because they considered it of too little importance, or whether they have not taken the pains to inform themselves on the subject, I am unable to say; but certain it is that nothing useful on the subject has been given in any scientific book on brewing, so far as I have been able to ascertain.

I have been at no little trouble in informing myself on this point, that I might be enabled to give the practical processes. In most of our large cities, immense quantities of these small beers, soda water, congress water, &c. &c. are consumed daily, and it is therefore a matter of great importance to know how they are made; from the fact that unless they are *properly* made, they will very seriously affect the health of those who consume them. I have gone into details on this subject, though, in another work prepared by me, entitled, "DETECTION OF FRAUD AND PROTECTION OF HEALTH," published in Philadelphia, to which the reader is referred for further information. One great object in preparing this work on brewing, is to give the necessary information on the subject that will enable those living in the interior of the country to carry through successfully all the processes of brewing.

In many small villages throughout the country, the subject under consideration must prove of great interest; for in those places, very often, a brewery of any size could not be supported; whereas the small beers soon to be described could be made and sold with a fair remuneration to the manufacturer. Again, these beers can be made for family use, and would be much better than the highly intoxicating liquors. The processes for making them are so simple, that every person can brew his own beer if he desires to do so; and it would be far preferable to purchasing beer of a *dishonest* dealer, for things prejudicial to health are too often used by them.

ROOT BEER.

THIS, though a cheap beer, is, with many persons, a great favourite; and it is thought by some to be quite conducive to health at certain seasons of the year. As I am treating of the practical processes of brewing, it would perhaps be out of place for me to speak of its good or bad qualities in this respect; but may be permitted to say, that if used in moderation, this beer is a wholesome beverage, when it is properly made. The sarsaparilla has a world-wide reputation at the present day, and if good in one form, such as "*syrup*," "*extract*," &c., it is undoubtedly so in the form of beer, as has been proved by observation. It is made as follows :—

Take of molasses 3 gallons; add to this 10 gallons of water at 60°. Let this stand for two hours, then pour into a barrel and add,

Powdered or bruised sassafras bark.........	½ pound
" or " wintergreen ".........	½ "
Bruised sarsaparilla root.....................	½ "
Yeast (fresh and good)......................	1 pint.

Water sufficient to fill the barrel, which is estimated to hold from 30 to 35 gallons is then put in. Let this ferment for twelve hours, when it can be bottled, if desired.

GINGER POP.

THIS is an article which has long been in use, but, owing to some cause, the proper mode of making it has been understood by only a few persons, who manufactured it on a very large scale, and endeavoured to blind the public by saying that there was great secrecy in making the genuine article. This is all "humbug." But be that as it may, if there is any great secret about it, or has been heretofore, it is all at an end now, for here is the process of making as good an article of "ginger pop" as any of them can make that are keeping our streets in a perfect uproar with their wagons, supplying the grocers and others. Their reason for trying to keep the information from the public on this point is, that they fear too many will be in the business, and the profits, of course, greatly reduced. It is made thus :—

Take of crushed (white) sugar.............. 28 pounds.
    Water .... 1 barrel.
    Yeast .... 1 pint.
    Powdered ginger (fine)................... 1 pound.
    Essence of lemon........................... ¼ ounce.
    Essence of cloves........................... ¼ ounce.

To the ginger pour half a gallon of boiling water, let stand for fifteen or twenty minutes; dissolve the sugar in two gallons of warm water. Pour each of these into a barrel half filled with cold water, then add the essences of lemon and cloves, and the yeast. Let stand for half an hour, when the barrel should be filled with cold water. When sufficiently fermented, it is poured into bottles and corked tightly until wanted for use. It wil. generally be sufficiently fermented in the course of a few hours.

MEAD.

MEAD is a very pleasant beverage, and is preferred by some persons to any other fermented drink. It has been observed that it is liable to disagree with certain individuals, which is generally attributed to the honey; but I have no doubt that the *real* cause in most cases is traceable to some mismanagement in making it, though it is known that honey is liable to produce colic when eaten in a crude state, or, that is, when first taken from the hive.

The following is the process for making mead :—

Take of honey three gallons; put this into a vessel and heat it to the boiling point, *taking great care that it does not boil over.* Pour this into a barrel half filled with water; let stand twenty or twenty-five minutes, then add,

Yeast..................................	1 pint.
Oil nutmeg...........................	1 tablespoonful.
Oil lemon or orange................	1 ounce.

Fill the barrel with water, and let stand for a short time, and it will ferment, after which it is fit for use.

THIS is made by taking of

Oil of spruce.................................. 40 drops.
Oil of sassafras............................... 40 "
Oil of wintergreen............................ 40 "

Pour one gallon of boiling water on the oils, then add,

Four gallons of cold water,
Three pints of molasses,
One pint of yeast.

Let stand for two hours, then bottle it for use.

BREWING OF CIDER.

CIDER is the result of fermentation of the expressed juice of different sorts of apples. Those varieties of apples which present a light ground, with a tinge of red streaks on the sunny side, have a smart acid flavour, and are aromatic and juicy, are said to be the best for making cider.

The process of brewing cider will be divided as follows :—

Gathering the fruit.
Grinding the apples.
Pressing the ground fruit.
Fermenting, racking, casking, bottling.

GATHERING THE FRUIT.

The fruit should be well ripe before it is taken from the tree; cider made with unripe fruit will be rough, harsh, and not so pleasant. The apples on a tree ripen at various intervals: they are thrown into heaps in the open air, where they remain till they become mellow, when they are fit for grinding. This is, however, an imperfect method. On a dry day in the gathering season, shake the limbs gently, and the ripest apples will fall off. This can be done from time to time as the fruit ripens. The yellow, or yellow mixed with red, should be separated from those with green pulps. The former are best adapted for making fine cider. The apples should be formed into heaps, ten inches or one foot thick, under a shelter, and so the air can circulate through them. Decayed or green fruit must not be ground up with good mellow fruit. Apples do not bear too great heat or cold,—they must be kept at a medium temperature.

Different sorts of fruits must be kept separate as much as possible,—sweet and sour apples must not be put together.

GRINDING THE FRUIT.

Each sort of apples should be ground separately—or such sorts together as ripen at the same time. The cider-mill is the apparatus now in general use for grinding the fruit. The action of the acids of the apple and pear, however, on the metal of the iron mill, produces cider of an unpleasant taste, and it should not be used. Lead must never be suffered to come in contact either with the ground fruit or the cider: it renders it poisonous. Cider-mills are of different constructions—some worked by hand, others by horse and water. The hand method reduces the fruit to the finest pulp, but the others perform the greatest amount of labour.

PRESSING THE GROUND FRUIT.

"Pommage" is the mass produced by grinding. This should not be pressed until it is ascertained, by a sample taken from the centre of the mass, that it has lost the luscious sweetness of taste, and gives to the nose a slight piquancy.

The pulp will generally be in a fit state for keeping in about twelve or sixteen hours. A heat will be generated, if it is kept longer, productive of premature fermentation. To remedy this, the pulp must be turned over often, so as to expose the interior of the mass to the external air.

The pommage being properly adapted, it is carried to the press, and a square cake or cheese formed of it : this is done by placing clean straw or reeds between layers of the pulp, or by putting the same into hair-cloths spread upon the vats, and placing them one upon another. Ten or twelve are thus arranged with perfect evenness, the square frame of the press being covered up also.

Upon the whole a strong board is placed, wider than the pile on which the boards rest. The straw or reeds should be frequently washed and dried, and should be

kept perfectly sweet, or the ill effects of their acidity will
be communicated to the cider. To this cake or cheese a
slight pressure should be applied, by lowering the screw
of the press. This pressure must be increased as the
cakes become dryer, until the juice is entirely expressed.
It is completed by the long lever and windlass. The
juice must then be strained through a coarse hair-
sieve, to keep back its grosser particles. The produce
must be put into proper vessels, and these may be either
open vats or close casks; but the best plan is to carry it
immediately from the press to the cask, thus saving the
inconvenience of or encumbrance of vats.

FERMENTATION, RACKING, AND CASKING.

Fermentation should not be conducted with too much
heat, or the process will be too rapid: nor should it be
carried on at too low a degree of temperature. Between
40° and 50° is the best heat for the production of good
cider. In cider brewing there is always a violent ebullition attending the fermentation; the bubbles rise, and
form a scum or crust over the surface of the liquor; the
ascent of additional fixed air breaks this crust,—another
is formed, and, in a short time, that crust is broken; and
thus the process continues. This effect gradually ceases,
and the fermentation proceeds less briskly. The liquor is
now clear, and has a piquant, vinous odour. If, when it
is in this state, there be any hissing sound heard in the
fermenting liquor, the room is too warm, and the cool external air must be admitted. The fermentation must here
be stopped; and to do this, rack off the clear liquor into

open vessels—keep it cool for a day or two—barrel it again, and place it in a cool place for the winter. Rack the cider by a small stream, and do not allow too much space between the tap and receiving-tub; thus avoiding a violent motion and another fermentation. Do not rack too often,—it reduces the strength. When the tendency to fermentation is great, do not fill the casks too full: the air on the surface of the liquor checks that process.

The casks into which cider is put, whenever racked off, should always have been thoroughly *scalded and dried*, and never be filled within a few gallons. The vessels should be filled up every two or three weeks, to supply the waste by the insensible fermentation, until the succeeding mash. Soon after the spring racking, but not till then, the casks may be gradually stopped, by first laying the cock on the bunghole, and in a few days forcing it in tightly, covering it with melted resin or other similar substance.

BOTTLING THE LIQUOR.

In the month of April, the cider will generally be in a fit state for bottling. Then, when the barometer is high and the wind from the northerly point, fill the bottles, and let them remain open until next morning. Then cork very tightly, secure with a string or wire, and cover the top of the cork with melted resin or wax.

CIDER CASKS.

The choice of proper vessels in which to keep the liquor after it has fermented, is a material point to be considered; .

15

for no liquor is so liable as cider to take the twang or taste of the cask: new vessels, it matters not how well the wood may have been seasoned, are apt to impart a disagreeable taste to liquors, and especially cider, unless due caution is previously used.

Scalding with hot water containing in solution a small quantity of salt, or with water in which some of the pommage has been boiled, and washing afterward with cider, rarely fail to correct this evil. Of all sorts of casks, beer vessels are the worst, as they always spoil cider,—and cider casks also spoil beer. Wine and brandy casks answer very well, if the tartar deposited on their sides be carefully scraped off and they be well scalded.

BURTON ALE.

FOR this, the richest malt is requisite,—such as will produce an extract of from eighty to eighty-five pounds per quarter gravity. Not more than one and a half or one and three-quarter barrel of this luscious ale should be drawn from a quarter. Heat of the first mash, 180°. Second mash, 190°. Third mash for return or table-beer. Hops of very superior quality,—from six to eight pounds per quarter. Boil the first mash one hour, and the second two hours: pitch at 58°, with eighteen pounds of yeast to every ten barrels. Attenuation must be carried on till reduced to twenty pounds per barrel, and the heat increased to 70° and 75°. Then skim and cleanse.

WINDSOR ALE.

THIS has been held in high estimation. The malt which is employed is ground, twenty-four or thirty-six hours before it is used. The strength is about two and a half barrels per quarter. The mashes are heated on the following scale :—

1st. 180° ; 2d. 190° ; 8d. 160°.

Boil the first wort briskly one hour, and the second two hours. Pitch the tun at 55°, and skim when the head assumes a close, yeasty appearance. Repeat the skimming until no yeast arises, and the gravity will be reduced to about twelve or fourteen pounds per barrel. Then start it into a vat, and add the hops : rouse them in until it is quite fine.

BAVARIAN BEER.

THE malt is first mixed with the water of ordinary temperature, in the Bavarian process. For one part of malt about 3·9 parts of water are employed. The whole is then allowed to rest for six or eight hours, after which the mashing is begun by mixing the mass with three parts of boiling water added gradually, during continual agitation, by which its temperature is raised to 106°. The thick part of the mash is then transferred to the copper, and heated to boiling during continual agitation ; and, after an hour's boiling, again returned to the mash-tun, and mixed thoroughly with its liquid contents,—by which the temperature in the mash-tun is raised to 133°.

The thick part of the mash is then once more transferred and boiled for an hour in the copper, and returned in the same way to the mash-tun,—by which the temperature is raised to 154°. The fluid part of the mash is next transferred by boiling or draining into the copper, boiled for a quarter of an hour, and then poured back on the mash in the tun, and mixed thoroughly with it. The temperature is hereby raised to from 167° to 180°. After agitation for a quarter of an hour, the mash is left at rest for one or one and a half hour, after which the clear wort is drawn off and treated in the usual manner.

By this method, it will be observed, the mashing is performed in the mash-tun at a temperature of from 133° to 180°, and the mass treated at a still higher temperature in the copper. At a temperature between 140° and 167°, starch is principally converted into sugar by diastase, while at still higher temperature into gum; which latter, therefore, must be formed in considerable quantity by this process,—the boiling of the liquid destroying the effect of the diastase in the solution, and preventing it from converting the gum subsequently into sugar. The albuminous matters are also coagulated by boiling, and clarify, by their precipitation, the worts, while the husk becomes much shrunk, and therefore retains but little of the worts; so that the latter drains off easily and more completely, and the second wort is therefore of only small avail. Notwithstanding the longer duration of the mashing, the mass is less apt to become sour, both on account of the temperature being above the degree favourable for this change, and more especially all albuminous matter, by which the conversion of starch, sugar, and gum into lactic acid is effected, is precipitated by the boiling.

The fermentation of the Bavarian beer is also peculiar, being performed very slowly, and at a very low temperature, (45° to 50°,) and mostly by the so-called lower or bottom fermentation, by which little or no acetic acid is formed, and all the gluten and other nitrogenized substances so completely precipitated, that, from the perfect absence of these substances, which, by their oxidation, form ferment for the acetic fermentation, Bavarian beer is capable of undergoing the acetous fermentation even by free exposure to the open air.

TABLE BEER, FROM BRAN AND SHORTS.

PROCESS as follows :—

Forty bushels of shorts.
Twenty bushels of bran.

Sixteen pounds of hops will give twenty-five barrels of small beer.

Boil your first copper; run into your mash-tun as much boiling water as, when reduced with cold, will bring it to the temperature of 150°; then commence your mashing operation, putting in two bushels of shorts and one bushel of bran at a time : when these are well mixed with the water, put in more, mash again, and so continue to do till all is in. It will take from half an hour to three-quarters to mash this quantity properly. Let the mash stand two hours; run down as in the preceding processes, and give your second liquor 165°; mash a second time; let stand one hour; boil the first worts one hour very briskly with half the hops, which should have been steeped, rubbed, and salted, as before directed; boil the second worts one hour and a half the same way, putting on the remainder of the hops, with one pound of ground mustard and five pounds of brown sugar, reduced by boiling to a colouring matter, as already directed. Make up your two boilings in your tun at the heat of 65°, giving three gallons of solid yeast; let the attenuation proceed ten degrees, or to 75°; then cleanse, and continue to fill the casks in the usual way.

It has been found that beer brewed from these materials has stood the summer heats much better than beer brewed from malt alone. This may be accounted for by the extract of malt possessing a much larger proportion of saccharine matter than that obtainable from bran and shorts. In families, this beer may be brewed in the proportion of one or two barrels at a time; and in the country, where brewer's yeast may not be procurable, leaven, diluted with blood-warm water, may be substituted for brewer's yeast, and will answer, but not so well; neither will attenuation go so high, as fermentation with leaven, when applied to liquids, is generally languid and slow.

GINGER WINE.

TAKE sixteen quarts of water; boil it; add one pound of bruised ginger; infuse it in the water for forty-eight hours, placed in a cask in some warm situation; after which time, strain off this liquor; add to it eight pounds of lump sugar, seven quarts of brandy, the juice of twelve lemons, and the rinds of twelve oranges; cut them; steep the fruit and the rinds of the oranges for twelve hours in the brandy; strain the brandy; add it to the other ingredients; bung up the cask; and in three or four weeks it will be fine: if it should not, a little dissolved isinglass will soon make it so.

SUBSTITUTE FOR BREWER'S YEAST.

TAKE six pounds of good malt and three gallons of boiling water; mash them well together; cover the mixture, and let it stand three hours; then draw off the liquor, and put two pounds of brown sugar to each gallon, stirring it well till the sugar is dissolved; then put into a cask just large enough to hold it, covering the bunghole with brown paper. Keep this cask in a temperature of ninety-eight degrees. Prepare the same quantity of malt and boiling water as before, but without sugar; then mix all together, and add one quart of yeast; let the cask stand open for forty-eight hours, and it will be fit for use. The quart of yeast should not be added to these two extracts at a higher heat than eighty degrees.

CURRANT WINE.

TAKE five gallons of currant-juice, and put it into a ten-gallon cask, with twenty pounds of Havanna or lump sugar; fill the cask with water; let it ferment with the bung out for some days; as it wastes, fill up with water; when done working, bung down; and in two or three months after, it will be fit for use. Two quarts of French brandy, added after the fermentation ceases, will improve the liquor, and communicate to it a preserving quality. Wine may be made from strawberries, raspberries, and cherries, in the same way.

SMALL BEER FOR SHIPPING.

12 bushels of pale malt.
12 bushels of amber malt.
___
24
___

14 pounds of hops.   24 barrels cleansed.
Let the malt be fine ground; first liquor, 172°; mash

one hour; stand one hour; run down smartly. Heat of second mash, 180°; mash one hour; stand two hours; boil two hours; making the length sufficiently long to give one barrel of beer. to each bushel of malt. Pitch the tun at 72°, giving one gallon of solid yeast; cleanse within twenty-four hours. The fresher this beer is sent out the better: being very thin in body, and low-priced, it cannot be expected to last long.

WELSH ALE.

THIS is a richly flavoured and luscious ale, and many persons are quite fond of it.

Process.

72 bushels of pale malt.
70 pounds of hops.
20 pounds of best brown sugar.
2 pounds of grains of paradise.

Heat of the first mashing liquor, 175°; mash one hour and a half, putting in the malt gradually, and mash unusually well, and let it stand two hours: second liquor at 190°; mash one hour and stand two more; run down as before; boil these two runs together for one hour and a half, putting in the hops, &c.; save the sugar, which is to be put in but a few minutes before striking off, at which time the rousing of the copper should commence, and so continue until the worts are nearly run off.

Small beer may be brewed, in the usual way, after both these worts, in which case cold water will answer as well as hot: pitch the strong worts at 62°, with a small proportion of good yeast, and let the fermenting heat rise to 80°: thus the attenuation will proceed eighteen degrees. Cleanse with salt and bean flour, as before directed, but in suitable proportion in point of quantity to the malt; fill in the usual way, and, when nearly done working, use fine ale to top with; before bunging down, putting into each barrel one large handful of scalded hops, that have been previously cooled down.

I must call the attention of the reader to the fact that the *grains of paradise*, spoken of in the process of making this ale, can be dispensed with, and the article should never be used. I have put it down, as it used with the other ingredients in making Welsh ale; but, for obvious reasons, I must condemn its use.

READING ALE.

THIS beer is made in a town named Reading, about thirty miles from London. It is in high repute, and has been for years. The process of making it is as follows:—

Scale of Brewing, suppose 22 barrels.

80 bushels of pale malt.
98 pounds of hops.
5 pounds coriander-seed, ground.
14 pounds best brown sugar.
4 pounds grains of paradise.

The malt should be several days ground, and if exposed on an open loft after grinding, so much the better. Boil the first copper; run on the mash-tun till you have your complement, then occasionally rouse the water with the mashing-oars, or dashers, till you get it down to 175°; put the malt in slowly; for fear of setting, keep mashing all the time, which should be continued full one hour; stand two hours; run the worts, when you set the tap, as fine as you can get them into the underback; this can be effected by drawing off, successively, five or six buckets of the first run, and throwing them over the grains in the mash-tun; when you perceive they come off glass-fine, lay by your bucket.

Give the second mashing liquor at 178°; mash three-quarters of an hour; stand one hour. Give the third liquor at 158°; mash half an hour; stand one hour; boil the first copper of worts,—which should take the half of the three runs,—one hour, as hard as you can; your second, two hours, in the same way; run the two boilings into one cooler, and pitch at 64°, giving one gallon of solid, smooth yeast: skim off the yeast, as in the case of Windsor ale, until the attenuation rises to 80°, which will have advanced it, from the pitching heat of 64°, sixteen degrees. Before you commence the operation of cleansing, mix one quarter of a pound of bay-salt with half a peck of malted bean flour; scatter this mixture over the surface of the tun; rouse well; cleanse, and fill in the usual way.

The grains of paradise can also be omitted in making this beer, and it will be much safer to do so.

WIRTEMBERG ALE.

PROCESS of brewing as follows:—

128 bushels of pale malt.
 32   "   " amber malt.
_____

160 bushels of malt.
_____

188 pounds of hops.
 28   "   honey.
 28   "   sugar.
  4   "   hartshorn shavings.
  4   "   coriander-seed, ground.
  1   "   caraway-seed, ground.

Cleanse 50 barrels ale.

Give the first mashing liquor at 172°; mash for one hour and a half; stand two hours; run down fine, but to some extent. Second mashing liquor, 180°; mash one hour; stand two hours; run down as before; get up the two worts; put in, with your hops, the other ingredients, save the honey and sugar, which are to be put into the copper but a few minutes before striking off, rousing the copper while any worts remain in it.

This ale should be boiled hard for one hour and a half; pitch the tun at 62°; raise the fermenting heat to 80°, which will generally rise in the course of 70 hours. Give of good solid yeast four gallons at first, and two gallons more in twelve hours after; rouse the tun each time.

HOCK.

In many parts of the world, but more especially in Germany, this beer has had a great run; but I cannot recommend its use, owing to the fact that *cocculus indicus* berries are used in the process. Wishing, however, to make this work as complete as possible, I will give the process of making this, as well as some other beers that I do not feel justified in saying are safe to use.

112 bushels of pale malt.
  48   "   amber malt.

160 bushels.

206 pounds of hops.
    4 " cocculus indicus berry, ground.
    2 " fabia amora, or bitter bean.
  20 " brown sugar, of good quality.
Cleanse 54 barrels.

First liquor, 176°; mash one hour and a quarter; stand one hour and a half; second liquor, 182°; mash one hour; stand two hours; when both worts are in the copper, and the hops and other ingredients, except the sugar, which is to be put in, as already directed, a little time before striking off; boil two hours and a quarter, as brisk as possible.

Pitch the tun at 64°, giving four gallons of solid yeast at once, and cleanse the second day, or in forty-eight hours; fill as before directed, and put into each barrel one handful of fresh steeped hops before bunging down.

SCURVY-GRASS ALE.·

FOR a long time, this species of ale has been considered a rectifier of the blood, and has been highly recommended by some medical men, as beneficial in certain conditions of the animal economy. It is generally thought to be quite wholesome and pleasant.

Process.

40 bushels of pale malt.
25 pounds of hops.
10   "   molasses.
2   "   Alexandria senna.   - ··
5 bushels of garden scurvy grass.

Cleanse 14 barrels of ale.

The malt should be fine ground; give the first liquor at 170°; stand one hour; heat of the second liquor, 172°; mash three-quarters of an hour; give the third mashing liquor at 160°; mash twenty minutes; stand half an hour; these three worts should be run into the copper together, and boil together for one hour gently, for one-quarter of an hour more as hard as you can; all the ingredients to be put in with the hops, except the molasses, which should only be put in a few minutes before striking off; from the time the molasses is put in, keep stirring the copper until its contents are nearly off.

About the middle of the fermentation, procure one pound of horse-radish, wash it well, dry it with a cloth, after which, slice it thin and throw it into the tun, rousing immediately after; when done, replace the tun-cover; pitch the worts at 66 degrees, with about two gallons of solid yeast; cleanse the third day, with the sweets on. This ale is drunk both hot and cold.

DORCESTER ALE.

WITH judicious management and the best materials, an article of ale bearing this name can be made that is equal to the best made in Edinburgh or London,—at least this is the opinion of good judges of ale.

    54 bushels of best pale malt.
    50 pounds of the best hops.
    1 pound of ginger, ground.
    ¼ " cinnamon.

Cleanse 14 barrels, reserving enough for filling.

Boil the copper; temper the liquor in the same to 185°; and, when ready, run it on your keeve a little at a time, putting in the malt and water gradually together, mashing at the same time: when the whole of the malt is thus got in, continue the operation of mashing half an hour; cap with dry malt, and let the mash stand one hour and a half.

Second liquor, 190°; mash three-quarters of an hour; stand two hours; in both mashes, get the worts as fine as you can into the underback; rub and salt, before mashing, 30 pounds of hops; infuse them in boiling water before mashing, and let the vessel containing them be close covered.

The other twenty pounds of hops should have been rubbed the evening before brewing, but not salted; put into another close vessel, covered with boiling water, and there suffered to digest for 12 hours; at the time of putting the hops in the copper, the extract in both cases is to be added; but the first 30 pounds in substance only to be added; these, with the two extracts, will be sufficient for the brewing; the remaining 20 pounds of hops will answer for single ale or table beer, but must be used on the same day.

The worts being now in the copper, with the hops and extract, boil briskly for one hour; after which, draw the fire, open the copper and ash-pit doors, and so let it stand one hour; then strike off gently on the cooler; when the worts are cooled down to 55°, prepare the puncheons, (suppose four,) containing four barrels each; see that they are dry, sweet, and clean; take three pints of solid yeast for each puncheon, to which add three quarts of worts, at 65°; mix and blend the worts and yeast together, putting this proportion to each cask containing four barrels; then fill up with the worts, at the heat of 55°, already spoken of; put in the tin workers, (one in each puncheon,) and when you perceive it begin to work freely, which probably will not be till the third or fourth day, begin to fill up the casks, and continue doing so from time to time, till they have finished working.

This mode of brewing appears to be particularly adapted for shipping to warm climates. The fermentation being slowly and coolly conducted, it is also well calculated for bottling.

Table beer may be made after this strong, of good quality, with cold water, if not overdrawn. Ten pounds of the steeped hops will be sufficient to preserve this beer; one hour's boiling will be enough: ferment as already directed, and add six pounds of sugar just before striking off, rousing, as before noticed, while any remains in the copper.

The annexed plate (fig. 16) shows the form and application of the worker, whether of tin or wood.

Fig. 16.

A, the cask in which the worker is placed.

B, the spout of the worker, which takes off the yeast.

C, the plug at the angle of the .worker, to admit the pipe of a tundish, in order to fill the cask as it works.

BOTTLING BEER.

UNDER proper management, this is a branch of trade that might be made very profitable; but if due care is not exercised it proves a heavy loss to the brewer. In consequence of bad management, some persons have lost, on an average, from two to three dozen bottles as well as beer, on every hop-head put up which happened to lie over till summer, or was bottled in that season. This loss is too heavy to expect much profit from a business so conducted. To obviate both these consequences, it is recommended that beer, ale, and porter intended for the bottle, should be carefully filtered through charcoal and sand, as directed in the operation of filtering (described further on.) Being thus purified from all its feculences and fermentable matter, it will be in the best possible state for taking the bottle, in that mild and gentle way that will not endanger the loss of one or the other. It will further have the good effect of recovering the beer or ale thus filtered from the flatness that will necessarily be induced by that operation, giving the liquor all the brisk-ness and activity that can be wished for. If beer, ale, or porter be intended for exportation to a warmer climate, the operation will be found particularly well suited to it.

Choose corks of the best quality, and steep them in pure, strong spirit, from the evening before you begin your bottling operation. This precaution is decidedly necessary to all beer intended to be shipped, or sent off to a warmer climate than our own, such as the East and West Indies, South America, &c.

In more temperate climates, the simple precaution of filtering alone will be found to answer every purpose, without steeping the corks in spirits. But suppose you bottle for home consumption : in that case you will naturally wish to have your beer, ale, and porter got up in the bottle in as short a space of time as possible; and you should pack away your bottles in dry straw in summer and in sawdust in winter, as the object at that season will be rather to accelerate than retard fermentation. Here you should carefully watch its progress from day to day, by drawing a bottle from the centre of the heap, as nearly as you can get it; place this bottle between you and the light, and if you perceive a chain of small bubbles in the neck of the bottle, immediately under the cork, you may conclude your beer is up in the bottle; then draw a few more bottles, and if the same appearance continues in them also, it is time to draw all your bottles from the heap they were originally packed in, and set them on their bottoms in a square frame ten inches deep,—size optional; fill up this frame with the bottles of porter or ale so drawn in a ripe state; then get one or more bushels of bay-salt, and scatter it as evenly as you can over the bottles, until the space between their necks is nearly half filled; then another course of bottles may be sunk between these, with their necks down through the salt, so as to form an upper tier; thus treated, not a single bottle will be found to break from the force of fermentation, and the salt will answer for a fresh supply of bottles as often as you may find it necessary to draw or send them out. This quantity will answer your purpose for years, if you only keep it dry. Another advantage, and no small one, derivable from a bottling operation conducted in this way, will be that a loft will be found more convenient for the purpose than a ground-floor, as less damp, and more likely to preserve the salt dry, which a more moist atmosphere would naturally dissolve.

The practice here recommended may, with equal success, be applied to cider and perry.

FILTERING OPERATION.

(See opposite Plate.)

IF the views which have long been held out of its effects on malt liquors, as well as other fermented liquors, be correct, this simple operation will do more toward their improvement and preservation than any thing yet attempted to be tried on them, after their fermentation has been completed; and for this plain reason, that it will at once disengage them from all fermentable matter, and render them transparently fine and preserving; thus immediately fitting them for the bottle or putting up into tight casks for home consumption or exportation, which will soon recover the beer or ale so treated from the flatness which will necessarily be induced by a long exposure to the air during the continuance of the operation; further to remedy which, it is recommended to put into each barrel, before the cask is filled with this beer, half a pound of ground rice; then fill; bung down tight; and in a

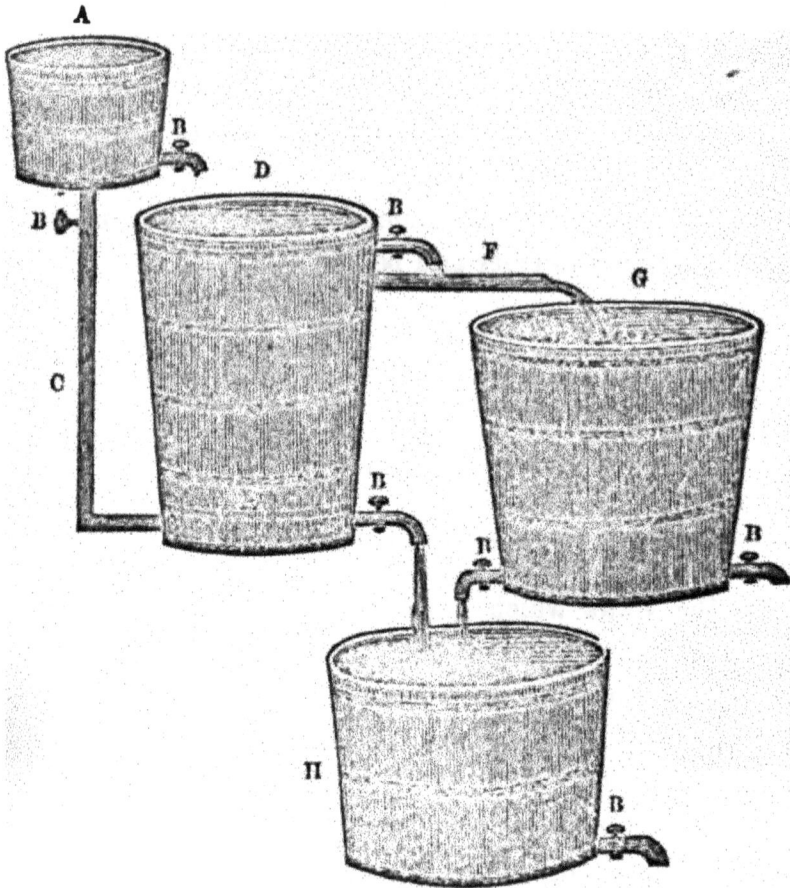

Fig. 17.

A, the fountain.

B B, the cocks.

C, the trunk communicating with the space between the two bottoms.

D, the filtering-tub.

E, the false bottom.

F, the spout for carrying off the ascending liquor.

G, the receiver of the filtered liquor by ascent.

H, the receiver of the filtered liquor by descent.

short time briskness and activity will be restored to the liquor, whether intended for draught or bottle.

This mode might, with equal success, be applied to every kind of fermented liquor, particularly to cider, wine, and perry; also to river and rain water.

There are two modes of filtration, one by descent the other by ascent; the latter operation seems to be the most perfect, though not the most economical or expeditious. The preparation of the filtering medium is as follows :—

The filtering vessel should be in proportion to the scale of work you intend operating on. The vessel containing the filter should have the form somewhat of an inverted cone, in proportion wider at top than bottom; over the bottom of this vessel should be placed a false one, about three or four inches distant from the other: this upper bottom should be perforated with holes, rather large bored, at the angles of every square inch of its surface; the false bottom being laid, provide two pieces of clean, thick blanketing, the full size of the vessel; lay these pieces one over the other; over them, a stratum six inches deep of rather coarsely powdered charcoal; this should be previously wetted with some of the beer or ale, till it is brought to the consistence of coarse mortar: over this lay another stratum of fine, clean pit sand, and so on, stratum upon stratum of sand and charcoal, till you have reached within six inches of the top; the cover of this vessel, which is also perforated with holes somewhat smaller than those of the bottom, is let down in the vessel to within one inch of the filtering medium, and in that position is well secured by buttons or otherwise.

When filtering by descent, you run the liquor over this

cover, which, by means of the holes, will be distributed evenly over the upper surface of the filter; and so you continue running on your liquor as fast as you see the operation will take it.

When you wish to filter by ascent, you introduce the liquor to be filtered between the two bottoms. As the fountain which supplies this liquor is higher than the filtering vessel, it will naturally force its way through the false bottom, filtering medium, &c., until it runs off pure at spout F into the receiver G.

Those persons who live on the banks or in the vicinity of our great rivers, such as the Mississippi, Missouri, Ohio, &c., may purify their drinking water in this way, with great advantage to their health and consequent increase of comfort to themselves and families. Having lived for several years on the Mississippi river, I can speak from experience, and must say the water for drinking and culinary purposes is horrible; but the process now under consideration would greatly improve it, and should be resorted to. It is also well adapted to those who navigate these waters, particularly such as proceed in steamboats, where convenient room can always be found for such useful and salutary purposes; and to them its use is strongly recommended.

CULTIVATION OF HOPS.

A RICH, deep soil, rather inclined to moisture, is, on the whole, the best adapted for the cultivation of hops; but it is observable that any soil (stiff clay only excepted) will suit the growing of hops, when properly prepared; and in many parts of Great Britain they use bog-land, which is fit for little else. The ground on which hops are to be planted should be made rich with that kind of manure best suited to the soil, and rendered fine and mellow by being ploughed deep, and harrowed several times. The hills should be at the distance of six or eight feet apart from each other, according to the richness of the soil. On lands that are rich, the vines will run the most; the hills must, therefore, be farther apart.

At the first opening of the spring, when the frosts are over, and vegetation begins, sets, or small pieces of the roots of hops, must be obtained from hops that are esteemed the best. Of the different kinds of hops, the long white is the most esteemed. It yields the greatest quantity, and is the most beautiful. The beauty of hops consists in their being of a pale bright-green colour. Care should be taken to obtain all of one sort; but if different sorts are used, they must be kept separate in the field, for there is a material difference in their time of ripening; and, if mixed in the field, will occasion extra trouble at the time of gathering them in.

Cut off from the main stalk or root, six inches in length, branches or suckers most healthy, and of the last year's growth, if possible to be procured; if not, they should be wrapped in cloth, and kept in a moist place, excluded from the air. A hole should then be made large and deep, and filled with rich, mellow earth. The sprouts should be set in this earth, with the bud upward, and the ground pressed close about them. If the buds have begun to open, the uppermost must be left just out of the ground; otherwise, cover it with the earth an inch. Two or three sets to a pole is sufficient; and three poles to a hill will be found most productive. Place one of the poles toward the north, the other two at equal distances, about two feet apart.

The sets are to be placed in the same manner as the poles, that they may the easier climb. The length of the poles may be from fourteen to eighteen feet, according as the soil is rich or poor. They should be so placed as to incline to each other, meet at their tops, and there be tied. This is contrary to the European method, but will be found best in America. In this way they will strengthen and support each other, and form so great a defence against the violent gusts of wind to which our climate is frequently subject in the months of July and August, as to prevent their being blown down.

They will, likewise, form a three-sided pyramid, which will have the greatest possible advantage from the sun. It is suggested by experience that hops which grow near the ground are the best. Too long poles, therefore, are not good, and care should be taken that the vines do not run beyond the poles: twisting off their tops will prevent it. The best kinds of wood for the poles are alder, ash, birch, elm, chestnut, and cedar: their durability is directly the

17

reverse of the order in which they stand. Charring or burning the end put in the ground, will preserve them.

Hops should not be poled until the spring of the second year, and then not till they have been dressed. All that is necessary for the first year, is to keep the hops free from weeds, and the ground light and mellow by hoeing and ploughing often, if the yard is large enough to admit of it. The vines, when run to the length of four or five feet, should be twisted together, to prevent their bearing the first year, for that would injure them.

In the month of March or April of the second year, the hills must be opened and all the sprouts or suckers cut off within one inch of the old root, but that must be left entire, with the roots that run down; then cover the hills with fine earth and manure. Hops must be dressed every year as soon as the frost will permit. On this being well done depends, in a great measure, the success of the crop. It is thought by many to be the best method to manure the hop-yard in the fall, and cover the hills entirely with the manure, asserting, with other advantages, that this prevents the frost from injuring the plants during the winter. Hops had better be gathered before they are full ripe than remain till they are over-ripe, for then they will lose their seed by the wind, or on being handled. The seed is the strongest part of the hop, and, when they get too ripe, will lose their green colour, which is very necessary to preserve the most valuable part of the plant.

The hops must be kept free from weeds, and the ground mellow by hoeing often through the season, and hills of earth gradually raised around the vines during the summer. The vines must be assisted in running on the poles with woollen yarn, suffering them to run with the sun. By the last of August, or the first of September, the hops will be ripe and fit to gather.

This may be easily known by their colour changing, and having a fragrant smell; their seed grows brown and hard. As soon as ripe, they must be gathered without delay, for a storm or frost will injure them materially. The most expeditious method of picking hops is to cut the vines three feet from the ground; pull up the poles and lay them on crotches, horizontally, at a height that may be conveniently got at; put under them a bin of equal length, and four may stand each side to pick at the same time.

Fair weather should always be chosen to gather hops, and they should never be gathered when dew or moisture is on them, as it subjects them to mould. They should be dried as soon as possible after they are gathered; if not immediately, they must be spread on a floor, to prevent their changing colour. The best mode of drying them is with a fire of charcoal and kiln, covered with hair-cloth, in the manner of a malt-kiln. Kilns covered with the splinters of walnut or ash will answer the purpose, and come much cheaper than hair-cloth. The fire should be steady and equal, and the hops gently stirred from time to time.

Great attention is necessary in this part of the business, that the hops be uniformly and sufficiently dried; if too much dried, they will look brown, and, of course, be materially injured in their quality, and proportionally reduced in price. If too little dried, they will lose their natural colour and flavour. They should be put on the hair.

cloth about six inches thick, after it has been moderately warmed; then a steady fire kept up till the hops are nearly dry, lest the moisture or sweat the fire has raised should fall back and change their colour.

After the hops have been in this situation seven, eight, or nine hours, and have got through sweating, and, when struck with a stick, will leap up, then throw them into a heap, mix them well, and spread them again, and let them remain till they are equally dry. While they are in a sweat, it will be best not to move them; for fear of burning, slacken the fire when the hops are to be turned, and increase it afterward.

Hops are sufficiently dried when their inner stalks break short, and their leaves become crisp and fall off easily. They will crackle a little when their seed is bursting, and then they should be removed from the kiln. Hops that are dried in the sun lose their rich flavour, and, if under cover, they are apt to ferment and change with the weather, and lose their strength: moderate fire preserves the colour and flavour of the hops, by evaporating the water, and retaining the oil of the hops.

After the hops are taken from the kiln, they should be laid in a heap, to acquire a little moisture to fit them for bagging. It would be well to exclude them from the air, by covering them with blankets. Three or four days will be sufficient for them to be in that state. When the hops are so moist that they may be pressed together without breaking, they are fit for bagging. Bags made of coarse linen cloth, eleven feet in length and seven in circumference, which hold about two hundred pounds weight, are most commonly used in Europe, but any size that suits best may be made use of.

To bag hops, a hole is made through the floor of the loft large enough for a man to pass through with ease. The bag must be fastened to a hoop larger than the hole, that the floor may serve to support the bag. For the convenience of handling the bags, some hops should be tied up in each corner, to serve as handles. The hops should be gradually thrown into the bag, and trod down continually, till it is filled. Its mouth must then be sown up, and the hops are fit for market.

The closer and harder hops are packed, the longer and better they will keep; but they should be kept dry.

In most parts of Great Britain where hops are culti-vated, they estimate the charge of cultivating one acre at forty-two dollars, for manuring and tilling, exclusive of poles and rent of land. Poles they estimate at sixteen dollars per annum, but in this country they would not amount to one-fourth that sum. One acre is computed to require three thousand poles, which will last from eight to twelve years, according to the quality of wood used.

The English growers of hops think they have a very indifferent crop if the produce of one acre does not amount to one hundred and thirty-three dollars; but much more frequently it amounts to two hundred dollars per acre.

In this country, experiments have been equally flatter-ing. A gentleman in the State of Massachusetts once raised hops, from one acre of ground, that sold for three hundred dollars. It is allowed that land in the State of New York is quite as favourable to the growth of hops.

Upon a low estimate, we may fairly compute the net profit of one acre of hops to be seventy dollars, over and above poles, manure, and labour; and, in a good year, more than this might reasonably be expected. There is one circumstance further that has some weight, and ought to be mentioned. In the English estimate, the expense put down is what they can hire the labour done for by those who make it their business to perform the different parts of the cultivation. A great saving may therefore be made by our farmers in the article of labour.

Added to this, the farmers here have another grand advantage, which is one of no small moment. The hop-harvest will come between our two great harvests, (the small grain and Indian corn,) without interfering with either; but in England the case is different,—the small grain and hop harvest come in together, and create a great scarcity of hands, it being then the most busy season of of the year.

It is found, by experience, that the soil and climate of the Eastern States are more favourable to the growth of hops than Great Britain, they not being so subject to moist, foggy weather of long continuance, which is very injurious to hops: and the Southern and Middle States are still more favourable to the growth of hops than the Eastern States, in point of flavour and strength.

The State of New York unites some advantages from either extreme of the Union. The cultivators of land in this State have every inducement which policy or interest can offer, to enter with spirit into the cultivation of hops;

as we shall thereby be able to supply our own demand, instead of sending to our neighbours for every bag we consume. But it is gratifying to say that but few hops are imported into this country at the present day, compared with the state of things a few years ago : and the time is near at hand when thousands of bags of hops will be shipped every season from our different seaports, to supply the wants of other nations.

THE END.

STEREOTYPED BY L. JOHNSON AND CO.
PHILADELPHIA.

PUBLICATIONS

OF

HENRY CAREY BAIRD,

SUCCESSOR TO E. L. CAREY,

No. 7 Hart's Building, Sixth Street above Chestnut, Philadelphia.

SCIENTIFIC AND PRACTICAL.

THE PRACTICAL MODEL CALCULATOR,

FOR the Engineer, Machinist, Manufacturer of Engine Work, Naval Architect, Miner, and Millwright. By OLIVER BYRNE, Compiler and Editor of the Dictionary of Machines, Mechanics, Engine Work and Engineering, and Author of various Mathematical and Mechanical Works. Illustrated by numerous Engravings. Now Complete, One large Volume, Octavo, of nearly six hundred pages...$3.50

It will contain such calculations as are met with and required in the Mechanical Arts, and establish models or standards to guide practical men. The Tables that are introduced, many of which are new, will greatly economise labour, and render the every-day calculations of the *practical man* comprehensive and easy. From every single calculation given in this work numerous other calculations are readily modelled, so that each may be considered the head of a numerous family of practical results.

The examples selected will be found appropriate, and in all cases taken from the actual practice of the present time. Every rule has been tested by the unerring results of mathematical research, and confirmed by experiment, when such was necessary.

The Practical Model Calculator will be found to fill a vacancy in the library of the practical working-man long considered a requirement. It will be found to excel all other works of a similar nature, from the great extent of its range, the exemplary nature of its well-selected examples, and from the easy, simple, and systematic manner in which the model calculations are established.

NORRIS'S HAND-BOOK FOR LOCOMOTIVE ENGINEERS AND MACHINISTS:

Comprising the Calculations for Constructing Locomotives. Manner of setting Valves, &c. &c. By SEPTIMUS NORRIS, Civil and Mechanical Engineer. In One Volume, 12mo., with illustrations... $1.50

With pleasure do we meet with such a work as Messrs. Norris and Baird have given us.—*Artisan.*

In this work, he has given what are called the "secrets of the business," in the rules to construct locomotives, in order that the million should be learned in all things.—*Scientific American.*

1

THE ARTS OF TANNING AND CURRYING

Theoretically and Practically considered in all their details. Being a full and comprehensive Treatise on the Manufacture of the various kinds of Leather. Illustrated by over two hundred Engravings. Edited from the French of De Fontenelle and Malapeyere. With numerous Emendations and Additions, by CAMPBELL MORFIT, Practical and Analytical Chemist. Complete in one Volume, octavo...$5.00

This important Treatise will be found to cover the whole field in the most masterly manner, and it is believed that in no other branch of applied science could more signal service be rendered to American Manufacturers.

The publisher is not aware that in any other work heretofore issued in this country, more space has been devoted to this subject than a single chapter; and in offering this volume to so large and intelligent a class as American Tanners and Leather Dressers, he feels confident of their substantial support and encouragement.

THE PRACTICAL COTTON-SPINNER AND MANU-FACTURER; Or, The Manager's and Overseer's Companion.

This works contains a Comprehensive System of Calculations for Mill Gearing and Machinery, from the first moving power through the different processes of Carding, Drawing, Slabbing, Roving, Spinning, and Weaving, adapted to American Machinery, Practice, and Usages. Compendious Tables of Yarns and Reeds are added. Illustrated by large Working-drawings of the most approved American Cotton Machinery. Complete in One Volume, octavo...$3.50

This edition of Scott's Cotton-Spinner, by OLIVER BYRNE, is designed for the American Operative. It will be found intensely practical, and will be of the greatest possible value to the Manager, Overseer, and Workman.

THE PRACTICAL METAL-WORKER'S ASSISTANT,

For Tin-Plate Workers, Brasiers, Coppersmiths, Zinc-Plate Ornamenters and Workers, Wire Workers, Whitesmiths, Blacksmiths, Bell Hangers, Jewellers, Silver and Gold Smiths, Electrotypers, and all other Workers in Alloys and Metals. By CHARLES HOLTZAPFFEL. Edited, with important additions, by OLIVER BYRNE. Complete in One Volume, octavo.............$4.00

It will treat of Casting, Founding, and Forging; of Tongs and other Tools; Degrees of Heat and Managemnet of Fires; Welding; of Heading and Swage Tools; of Punches and Anvils; of Hardening and Tempering; of Malleable Iron Castings. Case Hardening, Wrought and Cast Iron. The management and manipulation of Metals and Alloys, Melting and Mixing. The management of Furnaces, Casting and Founding with Metallic Moulds, Joining and Working Sheet Metal. Peculiarities of the different Tools employed. Processes dependent on the ductility of Metals. Wire Drawing, Drawing Metal Tubes, Soldering. The use of the Blowpipe, and every other known Metal-Worker's Tool. To the works of Holtzappfel, OLIVER BYRNE has added all that is useful and peculiar to the American Metal-Worker.

THE MANUFACTURE OF IRON IN ALL ITS VARIOUS BRANCHES:

To which is added an Essay on the Manufacture of Steel, by FREDERICK OVERMAN, Mining Engineer, with one hundred and fifty Wood Engravings. A new edition. In One Volume, octavo, five hundred pages..$5.00

We have now to announce the appearance of another valuable work on the subject which, in our humble opinion, supplies any deficiency which late improvements and discoveries may have caused, from the lapse of time since the date of "Mushet" and "Schrivenor." It is the production of one of our transatlantic brethren, Mr. Frederick Overman, Mining Engineer; and we do not hesitate to set it down as a work of great importance to all connected with the iron interest; one which, while it is sufficiently technological fully to explain chemical analysis, and the various phenomena of iron under different circumstances, to the satisfaction of the most fastidious, is written in that clear and comprehensive style as to be available to the capacity of the humblest mind, and consequently will be of much advantage to those works where the proprietors may see the desirability of placing it in the hands of their operatives.— *London Morning Journal.*

A TREATISE ON THE AMERICAN STEAM-ENGINE.

Illustrated by numerous Wood Cuts and other Engravings. By OLIVER BYRNE. In One Volume.   (In press.)

PROPELLERS AND STEAM NAVIGATION:

With Biographical Sketches of Early Inventors. By ROBERT MACFARLANE, C. E., Editor of the "Scientific American." In One Volume, 12mo. Illustrated by over Eighty Wood Engravings...75 cts.

The object of this "History of Propellers and Steam Navigation" is twofold. One is the arrangement and description of many devices which have been invented to propel vessels, in order to prevent many ingenious men from wasting their time, talents, and money on such projects. The immense amount of time, study, and money thrown away on such contrivances is beyond calculation. In this respect, it is hoped that it will be the means of doing some good.— *Preface.*

A TREATISE ON SCREW-PROPELLERS AND THEIR STEAM-ENGINES.

With Practical Rules and Examples by which to Calculate and Construct the same for any description of Vessels. By J. W. NYSTROM. Illustrated by twenty-five large working Drawings. In one Volume, octavo.

PRACTICAL SERIES.

THE AMERICAN MILLER AND MILLWRIGHT'S ASSIST-
ANT. $1.

THE TURNER'S COMPANION. 75 cts.

THE PAINTER, GILDER, AND VARNISHER'S COMPA-
NION. 75 cts.

THE DYER AND COLOUR-MAKER'S COMPANION. 75 cts.

THE BUILDER'S COMPANION. $1.

THE CABINET-MAKER'S COMPANION. 75 cts.

A TREATISE ON A BOX OF INSTRUMENTS. By THOMAS
KENTISH. $1.

THE PAPER-HANGER'S COMPANION. By J. ARROWSMITH.
75 cts.

THE ASSAYER'S GUIDE. By OSCAR M. LIEBER. 75 cts.

THE COMPLETE PRACTICAL BREWER. By M. L. BYRN. $1.

THE COMPLETE PRACTICAL DISTILLER. By M. L. BYRN. $1.

THE BOOKBINDER'S MANUAL.

THE PYROTECHNIST'S COMPANION. By G. W. MORTI-
MER. 75 cts.

WALKER'S ELECTROTYPE MANIPULATION. 75 cts.

THE AMERICAN MILLER AND MILLWRIGHT'S ASSISTANT:

By WILLIAM CARTER HUGHES, Editor of "The American Miller," (newspaper,) Buffalo, N. Y. Illustrated by Drawings of
the most approved Machinery. In One Volume, 12mo........$1

The author offers it as a substantial reference, instead of speculative theories,
which belong only to those not immediately attached to the business. Special
notice is also given of most of the essential improvements which have of late
been introduced for the benefit of the Miller.—*Savannah Republican.*

The whole business of making flour is most thoroughly treated by him.—
Bulletin

A very comprehensive view of the Millwright's business.—*Southern Literary
Messenger.*

THE TURNER'S COMPANION:

Containing Instructions in Concentric, Elliptic, and Eccentric
Turning. Also, various Plates of Chucks, Tools, and Instru-
ments, and Directions for using the Eccentric Cutter, Drill,
Vertical Cutter, and Circular Rest; with Patterns and Instruc-
tions for working them. Illustrated by numerous Engravings.
In One Volume, 12mo...75 cts.

The object of the Turner's Companion is to explain in a clear, concise, and
intelligible manner, the rudiments of this beautiful art.—*Savannah Republican.*

There is no description of turning or lathe-work that this elegant little treatise
does not describe and illustrate.— *Western Lit. Messenger.*

5

THE PAINTER, GILDER, AND VARNISHER'S COMPANION:

Containing Rules and Regulations for every thing relating to the arts of Painting, Gilding, Varnishing, and Glass Staining; numerous useful and valuable Receipts; Tests for the detection of Adulterations in Oils, Colours, &c., and a Statement of the Diseases and Accidents to which Painters, Gilders, and Varnishers are particularly liable; with the simplest methods of Prevention and Remedy. In one vol. small 12mo., cloth. 75cts.

Rejecting all that appeared foreign to the subject, the compiler has omitted nothing of real practical worth.—*Hunt's Merchant's Magazine.*

An excellent *practical work*, and one which the practical man cannot afford to be without.—*Farmer and Mechanic.*

It contains every thing that is of interest to persons engaged in this trade.—*Bulletin.*

This book will prove valuable to all whose business is in any way connected with painting.—*Scott's Weekly.*

Cannot fail to be useful.—*N. Y. Commercial.*

THE BUILDER'S POCKET COMPANION:

Containing the Elements of Building, Surveying, and Architecture; with Practical Rules and Instructions connected with the subject. By A. C. SMEATON, Civil Engineer, &c. In one volume, 12mo. $1.

CONTENTS:—The Builder, Carpenter, Joiner, Mason, Plasterer, Plumber, Painter, Smith, Practical Geometry, Surveyor, Cohesive Strength of Bodies, Architect.

It gives, in a small space, the most thorough directions to the builder, from the laying of a brick, or the felling of a tree, up to the most elaborate production of ornamental architecture. It is scientific, without being obscure and unintelligible, and every house-carpenter, master, journeyman, or apprentice, should have a copy at hand always.—*Evening Bulletin.*

Complete on the subjects of which it treats. A most useful practical work.—*Bult. American.*

It must be of great practical utility.—*Savannah Republican.*

To whatever branch of the art of building the reader may belong, he will find in this something valuable and calculated to assist his progress.—*Farmer and Mechanic.*

This is a valuable little volume, designed to assist the student in the acquisition of elementary knowledge, and will be found highly advantageous to every young man who has devoted himself to the interesting pursuits of which it treats.—*Va. Herald.*

1*

THE DYER AND COLOUR-MAKER'S COMPANION:

Containing upwards of two hundred Receipts for making Colors, on the most approved principles, for all the various styles and fabrics now in existence; with the Scouring Process, and plain Directions for Preparing, Washing-off, and Finishing the Goods. In one volume, small 12mo., cloth. 75 cts.

This is another of that most excellent class of practical books, which the publisher is giving to the public. Indeed we believe there is not, for manufacturers, a more valuable work, having been prepared for, and expressly adapted to their business.—*Farmer and Mechanic.*

It is a valuable book.—*Otsego Republican.*

We have shown it to some practical men, who all pronounced it the completest thing of the kind they had seen—*N. Y. Nation.*

THE CABINET-MAKER AND UPHOLSTERER'S COMPANION:

Comprising the Rudiments and Principles of Cabinet Making and Upholstery, with familiar instructions, illustrated by Examples, for attaining a proficiency in the Art of Drawing, as applicable to Cabinet Work; the processes of Veneering, Inlaying, and Buhl Work; the art of Dyeing and Staining Wood, Ivory, Bone, Tortoise-shell, etc. Directions for Lackering, Japanning, and Varnishing; to make French Polish; to prepare the best Glues, Cements, and Compositions, and a number of Receipts particularly useful for Workmen generally, with Explanatory and Illustrative Engravings. By J. STOKES. In one volume, 12mo., with illustrations. Second Edition. 75 cts.

THE PAPER-HANGER'S COMPANION:

In which the Practical Operations of the Trade are systematically laid down; with copious Directions Preparatory to Papering; Preventions against the effect of Damp in Walls; the various Cements and Pastes adapted to the several purposes of the Trade; Observations and Directions for the Panelling and Ornamenting of Rooms, &c. &c. By JAMES ARROWSMITH. In One Volume, 12mo. 75 cts.

THE ANALYTICAL CHEMIST'S ASSISTANT:

A Manual of Chemical Analysis, both Qualitative and Quantitative, of Natural and Artificial Inorganic Compounds; to which are appended the Rules for Detecting Arsenic in a Case of Poisoning. By FREDERIK WŒHLER, Professor of Chemistry in the University of Göttingen. Translated from the German, with an Introduction, Illustrations, and copious Additions, by OSCAR M. LIEBER. Author of the "Assayer's Guide." In one Volume, 12mo. $1.25.

———

RURAL CHEMISTRY:

An Elementary Introduction to the Study of the Science, in its relation to Agriculture and the Arts of Life. By EDWARD SOLLEY, Professor of Chemistry in the Horticultural Society of London. From the Third Improved London Edition. 12mo. $1.25.

———

THE FRUIT, FLOWER, AND KITCHEN GARDEN.

By PATRICK NEILL, L.L.D.

Thoroughly revised, and adapted to the climate and seasons of the United States, by a Practical Horticulturist. Illustrated by numerous Engravings. In one volume, 12mo. $1.25.

———

HOUSEHOLD SURGERY; OR, HINTS ON EMERGENCIES.

By J. F. SOUTH, one of the Surgeons of St. Thomas's Hospital. In one volume, 12mo. Illustrated by nearly fifty Engravings. $1.25.

———

HOUSEHOLD MEDICINE.

In one volume, 12mo. Uniform with, and a companion to, the above. (In immediate preparation.)

THE COMPLETE PRACTICAL BREWER;

Or, Plain, Concise, and Accurate Instructions in the Art of Brewing Beer, Ale, Porter, &c. &c., and the Process of Making all the Small Beers. By M. LAFAYETTE BYRN, M. D. With Illustrations, 12mo. $1.

THE COMPLETE PRACTICAL DISTILLER;

By M. LAFAYETTE BYRN, M. D. With Illustrations, 12mo. $1.

THE ENCYCLOPEDIA OF CHEMISTRY, PRACTI-CAL AND THEORETICAL:

Embracing its application to the Arts, Metallurgy, Mineralogy, Geology, Medicine, and Pharmacy. By JAMES C. BOOTH, Melter and Refiner in the United States Mint; Professor of Applied Chemistry in the Franklin Institute, etc.; assisted by CAMPBELL MORFIT, author of "Chemical Manipulations," etc. Complete in one volume, royal octavo, 978 pages, with numerous wood cuts and other illustrations. $5.

It covers the whole field of Chemistry as applied to Arts and Sciences. * * * As no library is complete without a common dictionary, it is also our opinion that none can be without this Encyclopedia of Chemistry.—*Scientific American.*

A work of time and labour, and a treasury of chemical information.—*North American.*

By far the best manual of the kind which has been presented to the American public.—*Boston Courier.*

PERFUMERY; ITS MANUFACTURE AND USE:

With Instructions in every branch of the Art, and Receipts for all the Fashionable Preparations; the whole forming a valuable aid to the Perfumer, Druggist, and Soap Manufacturer. Illustrated by numerous Wood-cuts. From the French of Celnart, and other late authorities. With Additions and Improvements by CAMPBELL MORFIT, one of the Editors of the "Encyclopedia of Chemistry." In one volume, 12mo., cloth. $1.

A TREATISE ON A BOX OF INSTRUMENTS,

And the SLIDE RULE, with the Theory of Trigonometry and Logarithms, including Practical Geometry, Surveying, Measuring of Timber, Cask and Malt Gauging, Heights and Distances. By THOMAS KENTISH. In One Volume, 12mo. $1.

STEAM FOR THE MILLION.

An Elementary Outline Treatise on the Nature and Management of Steam, and the Principles and Arrangement of the Engine. Adapted for Popular Instruction, for Apprentices, and for the use of the Navigator. With an Appendix containing Notes on Expansive Steam, &c. In One Volume, 8vo...37½ cts.

SYLLABUS OF A COMPLETE COURSE OF LECTURES ON CHEMISTRY:

Including its Application to the Arts, Agriculture, and Mining, prepared for the use of the Gentlemen Cadets at the Hon. E. I. Co.'s Military Seminary, Addiscombe. By Professor E. SOLLY, Lecturer on Chemistry in the Hon. E. I. Co.'s Military Seminary. Revised by the Author of "Chemical Manipulations." In one volume, octavo, cloth. $1.25.

THE ASSAYER'S GUIDE;

Or, Practical Directions to Assayers, Miners, and Smelters, for the Tests and Assays, by Heat and by Wet Processes, of the Ores of all the principal Metals, and of Gold and Silver Coins and Alloys. By OSCAR M. LIEBER, late Geologist to the State of Mississippi. 12mo. With Illustrations. 75 cts.

THE BOOKBINDER'S MANUAL.

Complete in one Volume, 12mo.

ELECTROTYPE MANIPULATION:

Being the Theory and Plain Instructions in the Art of Working in Metals, by Precipitating them from their Solutions, through the agency of Galvanic or Voltaic Electricity. By CHARLES V. WALKER, Hon. Secretary to the London Electrical Society, etc Illustrated by Wood-cuts. A New Edition, from the Twenty-fifth London Edition. 12mo. 75 cts.

PHOTOGENIC MANIPULATION:

Containing the Theory and Plain Instructions in the Art of Photography, or the Productions of Pictures through the Agency of Light; including Calotype, Chrysotype, Cyanotype, Chroma-type, Energiatype, Anthotype, Amphitype, Daguerreotype, Thermography, Electrical and Galvanic Impressions. By GEORGE THOMAS FISHER, Jr., Assistant in the Laboratory of the London Institution. Illustrated by wood-cuts. In one volume, 24mo., cloth. 62 cts.

MATHEMATICS FOR PRACTICAL MEN:

Being a Common-Place Book of Principles, Theorems, Rules, and Tables, in various departments of Pure and Mixed Mathematics, with their Applications; especially to the pursuits of Surveyors, Architects, Mechanics, and Civil Engineers, with numerous Engravings. By OLINTHUS GREGORY, L. L. D. $1.50.

Only let men awake, and fix their eyes, one while on the nature of things, another while on the application of them to the use and service of mankind. —*Lord Bacon.*

ELEMENTARY COURSE OF INSTRUCTION ON ORDNANCE, GUNNERY, AND STEAM;

Prepared for the use of the Midshipmen at the Naval School. By JAMES H. WARD, U. S. N. In one Volume, octavo. $2.50.

SHEEP-HUSBANDRY IN THE SOUTH:

Comprising a Treatise on the Acclimation of Sheep in the Southern States, and an Account of the different Breeds. Also, a Complete Manual of Breeding, Summer and Winter Management, and of the Treatment of Diseases. With Portraits and other Illustrations. By HENRY S. RANDALL. In One Volume, octavo...$1.25

ELWOOD'S GRAIN TABLES:

Showing the value of Bushels and Pounds of different kinds of Grain, calculated in Federal Money, so arranged as to exhibit upon a single page the value at a given price from *ten cents to two dollars* per bushel, of any quantity from *one pound to ten thousand bushels.* By J. L. ELWOOD. A new Edition. In One Volume, 12mo...$1

To Millers and Produce Dealers this work is pronounced by all who have it in use, to be superior in arrangement to any work of the kind published—and *unerring accuracy in every calculation may be relied upon in every instance.*
☞ A reward of Twenty-five Dollars is offered for an error of one cent found in the work.

MISS LESLIE'S COMPLETE COOKERY.

Directions for Cookery, in its Various Branches. By MISS LESLIE. Forty-second Edition. Thoroughly Revised, with the Addition of New Receipts. In One Volume, 12mo, half bound, or in sheep...$1

In preparing a new and carefully revised edition of this my first work on cookery, I have introduced improvements, corrected errors, and added new receipts, that I trust will on trial be found satisfactory. The success of the book (proved by its immense and increasing circulation) affords conclusive evidence that it has obtained the approbation of a large number of my countrywomen; many of whom have informed me that it has made practical housewives of young ladies who have entered into married life with no other acquirements than a few showy accomplishments. Gentlemen, also, have told me of great improvements in the family table, after presenting their wives with this manual of domestic cookery, and that, after a morning devoted to the fatigues of business, they no longer find themselves subjected to the annoyance of an ill-dressed dinner.—*Preface.*

MISS LESLIE'S TWO HUNDRED RECEIPTS IN FRENCH COOKERY.

A new Edition, in cloth.. ....25 cts.

TABLES OF LOGARITHMS FOR ENGINEERS AND MACHINISTS:

Containing the Logarithms of the Natural Numbers, from 1 to ,00000, by the help of Proportional Differences. And Logarithmic Sines, Cosines, Tangents, Co-tangents, Secants, and Cosecants, for every Degree and Minute in the Quadrant. To which are added, Differences for every 100 Seconds. By OLIVER BYRNE, Civil, Military, and Mechanical Engineer. In One Volume, 8vo. cloth..........$1

TWO HUNDRED DESIGNS FOR COTTAGES AND VILLAS, &c. &c.

Original and Selected. By THOMAS U. WALTER, Architect of Girard College, and JOHN JAY SMITH, Librarian of the Philadelphia Library. In Four Parts, quarto..........$10

ELEMENTARY PRINCIPLES OF CARPENTRY.

By THOMAS TREDGOLD. In One Volume, quarto, with numerous Illustrations..........$2.50

A TREATISE ON BREWING AND DISTILLING,

In One Volume, 8vo. (In press.)

FAMILY ENCYCLOPEDIA

Of Useful Knowledge and General Literature; containing about Four Thousand Articles upon Scientific and Popular Subjects. With Plates. By JOHN L. BLAKE, D. D. In One Volume, 8vo, cloth extra..........$3.50

SYSTEMATIC ARRANGEMENT OF COKE'S FIRST INSTITUTES OF THE LAWS OF ENGLAND.

By J. H. THOMAS. Three Volumes, 8vo, law sheep..........$12

THE PYROTECHNIST'S COMPANION;

Or, A Familiar System of Recreative Fire-Works. By G. W. MORTIMER. Illustrated by numerous Engravings. 12mo. 75 cts.

STANDARD ILLUSTRATED POETRY.

THE TALES AND POEMS OF LORD BYRON:

Illustrated by HENRY WARREN. In One Volume, royal 8vo. with 10 Plates, scarlet cloth, gilt edges..................................$5
Morocco extra..$7

It is illustrated by several elegant engravings, from original designs by WARREN, and is a most splendid work for the parlour or study.—*Boston Evening Gazette.*

CHILDE HAROLD; A ROMAUNT BY LORD BYRON:

Illustrated by 12 Splendid Plates, by WARREN and others. In One Volume, royal 8vo., cloth extra, gilt edges.....................$5
Morocco extra $7

Printed in elegant style, with splendid pictures, far superior to any thing of the sort usually found in books of this kind.—*N. Y. Courier.*

THE FEMALE POETS OF AMERICA.

By RUFUS W. GRISWOLD. A new Edition. In One Volume, royal 8vo. Cloth, gilt...$2.50
Cloth extra, gilt edges...$3
Morocco super extra $4.50

The best production which has yet come from the pen of Dr. GRISWOLD, and the most valuable contribution which he has ever made to the literary celebrity of the country.—*N. Y. Tribune.*

THE LADY OF THE LAKE:

By SIR WALTER SCOTT. Illustrated with 10 Plates, by CORBOULD and MEADOWS. In One Volume, royal 8vo. Bound in cloth extra, gilt edges..$5
Turkey morocco super extra..$7

This is one of the most truly beautiful books which has ever issued from the American press.

LALLA ROOKH; A ROMANCE BY THOMAS MOORE:

Illustrated by 13 Plates, from Designs by CORBOULD, MEADOWS, and STEPHANOFF. In One Volume, royal 8vo. Bound in cloth extra, gilt edges..$5
Turkey morocco super extra..$7

This is published in a style uniform with the "Lady of the Lake."

THE POETICAL WORKS OF THOMAS GRAY:

With Illustrations by C. W. Radcliff. Edited with a Memoir, by Henry Reed, Professor of English Literature in the University of Pennsylvania. In One Volume, 8vo. Bound in cloth extra, gilt edges..$3.50
Turkey morocco super extra...$5.50

It is many a day since we have seen issued from the press of our country a volume so complete and truly elegant in every respect. The typography is faultless, the illustrations superior, and the binding superb.—*Troy Whig.*

We have not seen a specimen of typographical luxury from the American press which can surpass this volume in choice elegance.—*Boston Courier.*

It is eminently calculated to consecrate among American readers, (if they have not been consecrated already in their hearts,) the pure, the elegant, the refined, and, in many respects, the sublime imaginings of Thomas Gray.—*Richmond Whig.*

THE POETICAL WORKS OF HENRY WADSWORTH LONGFELLOW:

Illustrated by 10 Plates, after Designs by D. Huntingdon, with a Portrait. Ninth Edition. In One Volume, royal 8vo. Bound in cloth extra, gilt edges.....................................$5
Morocco super extra...$7

This is the very lu .ury of literature—Longfellow's charming poems presented in a form of unsurpassed beauty.—*Neal's Gazette.*

POETS AND POETRY OF ENGLAND IN THE NINETEENTH CENTURY.

By Rufus W. Griswold. Illustrated. In One Volume, royal 8vo. Bound in cloth..$3
Cloth extra, gilt edges...$3.50
Morocco super extra..$5

Such is the critical acumen discovered in these selections, that scarcely a page is to be found but is redolent with beauties, and the volume itself may be regarded as a galaxy of literary pearls.—*Democratic Review.*

THE TASK, AND OTHER POEMS.

By William Cowper. Illustrated by 10 Steel Engravings. In One Volume, 12mo. Cloth extra, gilt edges...................$2
Morocco extra...$2

"The illustrations in this edition of Cowper are most exquisitely designed and engraved."

THE FEMALE POETS OF GREAT BRITAIN.

With Copious Selections and Critical Remarks. By FREDERIC ROWTON. With Additions by an American Editor, and finely engraved Illustrations by celebrated Artists. In One Volume, royal 8vo. Bound in cloth extra, gilt edges.........$5
Turkey morocco............$7

Mr. Rowton has presented us with admirably selected specimens of nearly one hundred of the most celebrated female poets of Great Britain, from the time of Lady Juliana Berners, the first of whom there is any record, to the Mitfords, the Hewitts, the Cooks, the Barretts, and others of the present day.—*Hunt's Merchants' Magazine.*

SPECIMENS OF THE BRITISH POETS.

From the time of Chaucer to the end of the Eighteenth Century. By THOMAS CAMPBELL. In One Volume, royal 8vo. (In press.)

THE POETS AND POETRY OF THE ANCIENTS:

By WILLIAM PETER, A. M. Comprising Translations and Specimens of the Poets of Greece and Rome, with an elegant engraved View of the Coliseum at Rome. Bound in cloth......$3
Cloth extra, gilt edges.........$3.50
Turkey morocco super extra.........$5

It is without fear that we say that no such excellent or complete collection has ever been made. It is made with skill, taste, and judgment.—*Charleston Patriot.*

THE POETICAL WORKS OF N. PARKER WILLIS.

Illustrated by 16 Plates, after designs by E. LEUTZE. In One Volume, royal 8vo. A new Edition. Bound in cloth extra, gilt edges.........$5
Turkey morocco super extra.........$7

This is one of the most beautiful works ever published in this country.—*Courier and Inquirer.*

Pure and perfect in sentiment, often in expression, and many a heart has been won from sorrow or roused from apathy by his earlier melodies. The illustrations are by LEUTZE,—a sufficient guarantee for their beauty and grace. As for the typographical execution of the volume, it will bear comparison with any English book, and quite surpasses most issues in America.—*Neal's Gazette.*

The admirers of the poet could not have his gems in a better form for holiday presents.—*W. Continent.*

MISCELLANEOUS.

JOURNAL OF ARNOLD'S EXPEDITION TO QUEBEC, IN 1775.

By Isaac Senter, M. D.   8vo, boards.................62 cts.

ADVENTURES OF CAPTAIN SIMON SUGGS;

And other Sketches. By Johnson J. Hooper. With Illustrations. 12mo, paper...50 cts.
Cloth...75 cts.

AUNT PATTY'S SCRAP-BAG.

By Mrs. Caroline Lee Hentz, Author of "Linda." 12mo.
Paper covers...50 cts.
Cloth...75 cts.

BIG BEAR OF ARKANSAS;

And other Western Sketches. Edited by W. T. Porter. In
One Volume, 12mo, paper.....................................50 cts.
Cloth...75 cts.

COMIC BLACKSTONE.

By Gilbert Abbot a' Becket. Illustrated. Complete in One
Volume. Cloth...75 cts.

GHOST STORIES.

Illustrated by Designs by Darley. In One Volume, 12mo,
paper covers...50 cts.

MODERN CHIVALRY; OR, THE ADVENTURES OF CAPTAIN FARRAGO AND TEAGUE O'REGAN.

By H. H. Brackenridge. Second Edition since the Author's
death. With a Biographical Notice, a Critical Disquisition on
the Work, and Explanatory Notes. With Illustrations, from
Original Designs by Darley. Two volumes, paper covers 75 cts.
Cloth or sheep..$1.00

COMPLETE WORKS OF LORD BOLINGBROKE:

With a Life, prepared expressly for this Edition, containing Additional Information relative to his Personal and Public Character, selected from the best authorities. In Four Volumes, 8vo. Bound in cloth...$7.00
In sheep...8.00

CHRONICLES OF PINEVILLE.

By the Author of "Major Jones's Courtship." Illustrated by DARLEY. 12mo, paper...50 cts.
Cloth...75 cts.

GILBERT GURNEY.

By THEODORE HOOK. With Illustrations. In One Volume, 8vo., paper...50 cts.

MEMOIRS OF THE GENERALS, COMMODORES, AND OTHER COMMANDERS,

Who distinguished themselves in the American Army and Navy, during the War of the Revolution, the War with France, that with Tripoli, and the War of 1812, and who were presented with Medals, by Congress, for their gallant services. By THOMAS WYATT, A. M., Author of "History of the Kings of France." Illustrated with Eighty-two Engravings from the Medals. 8vo. Cloth gilt...$2.00
Half morocco...$2.50

GEMS OF THE BRITISH POETS.

By S. C. HALL. In One Volume, 12mo., cloth.............$1.00
Cloth, gilt..$1.25

VISITS TO REMARKABLE PLACES:

Old Halls, Battle Fields, and Scenes Illustrative of striking passages in English History and Poetry. By WILLIAM HOWITT. In Two Volumes, 8vo, cloth....................................$4.00

2*

NARRATIVE OF THE ARCTIC LAND EXPEDITION.

By CAPTAIN BACK, R. N. In One Volume, 8vo, boards...$2.00

THE MISCELLANEOUS WORKS OF WILLIAM HAZLITT.

Including Table-talk; Opinions of Books, Men, and Things; Lectures on Dramatic Literature of the Age of Elizabeth; Lectures on the English Comic Writers; The Spirit of the Age, or Contemporary Portraits. Five Volumes, 12mo., cloth......$5.00
Half calf..$6.25

FLORAL OFFERING.

A Token of Friendship. Edited by FRANCES S. OSGOOD. Illustrated by 10 beautiful Bouquets of Flowers. In One Volume, 4to, muslin, gilt edges................................$3.50
Turkey morocco super extra...........................$5.50

THE HISTORICAL ESSAYS,

Published under the title of "Dix Ans D'Etude Historique," and Narratives of the Merovingian Era; or, Scenes in the Sixth Century. With an Autobiographical Preface. By AUGUSTUS THIERRY, Author of the "History of the Conquest of England by the Normans." 8vo., paper.............................$1.00
Cloth $1.25

BOOK OF THE SEASONS;

Or, The Calendar of Nature. By WILLIAM HOWITT. One Volume, 12mo, cloth...................................$1
Calf extra...$2

PICKINGS FROM THE "PORTFOLIO OF THE REPORTER OF THE NEW ORLEANS PICAYUNE."

Comprising Sketches of the Eastern Yankee, the Western Hoosier, and such others as make up Society in the great Metropolis of the South. With Designs by DARLEY. 18mo., paper...50 cts.
Cloth...75 cts.

NOTES OF A TRAVELLER

On the Social and Political State of France, Prussia, Switzerland, Italy, and other parts of Europe, during the present Century. By SAMUEL LAING. In One Volume, 8vo., cloth........$1

HISTORY OF THE CAPTIVITY OF NAPOLEON AT ST. HELENA.

By GENERAL COUNT MONTHOLON, the Emperor's Companion in Exile and Testamentary Executor. One Volume, 8vo., cloth, $2.50
Half morocco..$3.00

MY SHOOTING BOX.

By FRANK FORRESTER, (Henry Wm. Herbert, Esq.,) Author of "Warwick Woodlands," &c. With Illustrations, by DARLEY. One Volume, 12mo., cloth.......................................75 cts.
Paper covers...50 cts.

MYSTERIES OF THE BACKWOODS:

Or, Sketches of the South-west—including Character, Scenery, and Rural Sports. By T. B. THORPE, Author of "Tom Owen, the Bee-Hunter," &c. Illustrated by DARLEY. 12mo, cloth, 75 cts.
Paper.. 50 cts.

NARRATIVE OF THE LATE EXPEDITION TO THE DEAD SEA.

From a Diary by one of the Party. Edited by EDWARD P. MONTAGUE. 12mo, cloth...$1

MY DREAMS:

A Collection of Poems. By Mrs. LOUISA S. McCORD. 12mo, boards ....75 cts.

AMERICAN COMEDIES.

By JAMES K. PAULDING and WM. IRVING PAULDING. One Volume, 16mo, boards...50 cts.

RAMBLES IN YUCATAN;

Or, Notes of Travel through the Peninsula: including a Visit to the Remarkable Ruins of Chi-chen, Kabah, Zayi, and Uxmal. With numerous Illustrations. By B. M. NORMAN. Seventh Edition. In One Volume, octavo, cloth.............................$2

THE AMERICAN IN PARIS.

By JOHN SANDERSON. A New Edition. In Two Volumes, 12mo, cloth.......:..$1.50

This is the most animated, graceful, and intelligent sketch of French manners, or any other, that we have had for these twenty years.—*London Monthly Magazine.*

ROBINSON CRUSOE.

A Complete Edition, with Six Illustrations. One Volume, 8vo, paper covers..$1.00
Cloth, gilt edges...$1.25

SCENES IN THE ROCKY MOUNTAINS,

And in Oregon, California, New Mexico, Texas, and the Grand Prairies; or, Notes by the Way. By RUFUS B. SAGE. Second Edition. One Volume, 12mo, paper covers ....50 cts.
With a Map, bound in cloth..75 cts.

THE PUBLIC MEN OF THE REVOLUTION:

Including Events from the Peace of 1783 to the Peace of 1815. In a Series of Letters. By the late Hon. WM. SULLIVAN, LL. D. With a Biographical Sketch of the Author, by his son, JOHN T. S. SULLIVAN. With a Portrait. In One Volume, 8vo. cloth $2.50

ACHIEVEMENTS OF THE KNIGHTS OF MALTA.

By ALEXANDER SUTHERLAND. In One Volume, 16mo, cloth, $1.00
Paper...75 cts.

ATALANTIS.

A Poem. By WILLIAM GILMORE SIMMS. 12mo, boards, 50 cts.

LIVES OF MEN OF LETTERS AND SCIENCE.

By HENRY LORD BROUGHAM. Two Volumes, 12mo, cloth, $1.50
Paper $1.00

THE LIFE, LETTERS, AND JOURNALS OF LORD BYRON.

By THOMAS MOORE. Two Volumes, 12mo, cloth..................$2

THE BOWL OF PUNCH.

Illustrated by Numerous Plates. 12mo, paper............50 cts.

CHILDREN IN THE WOOD.

Illustrated by HARVEY. 12mo, cloth, gilt.....................50 cts.
Paper ....25 cts.

TOWNSEND'S NARRATIVE OF THE BATTLE OF BRANDYWINE.

One Volume, 8vo, boards...$1.

THE POEMS OF C. P. CRANCH.

In One Volume, 12mo, boards............................87 cts.

THE WORKS OF BENJ. DISRAELI.

Two Volumes, 8vo, cloth $2
Paper covers..$1

NATURE DISPLAYED IN HER MODE OF TEACHING FRENCH.

By N. G. DUFIEF. Two Volumes, 8vo, boards...................$5

NATURE DISPLAYED IN HER MODE OF TEACHING SPANISH.

By N. G. DUFIEF. In Two Volumes, 8vo, boards...............$7

FRENCH AND ENGLISH DICTIONARY.

By N. G. Dufief. In One Volume, 8vo, sheep...................$5

FROISSART BALLADS AND OTHER POEMS.

By Philip Pendleton Cooke. In One Volume, 12mo, boards..50 cts.

THE LIFE OF RICHARD THE THIRD.

By Miss Halsted. In One Volume, 8vo, cloth............$1.50

THE LIFE OF NAPOLEON BONAPARTE.

By William Hazlitt. In Three Volumes, 12mo, cloth......$3
Half calf...$4

TRAVELS IN GERMANY, BY W. HOWITT.
EYRE'S NARRATIVE.   BURNE'S CABOOL.

In One Volume, 8vo, cloth..$1.25

CAMPANIUS HOLMES'S ACCOUNT OF NEW SWEDEN.

8vo, boards...$1.50

IMAGE OF HIS FATHER.

By Mayhew. Complete in One Volume, 8vo, paper....25 cts.

SPECIMENS OF THE BRITISH CRITICS.

By Christopher North (Professor Wilson). 12mo, cloth. $1.00

A TOUR TO THE RIVER SAUGENAY, IN LOWER CANADA.

By Charles Lanman. In One Volume, 16mo, cloth....62 cts
Paper.. 50 cts

TRAVELS IN AUSTRIA, RUSSIA, SCOTLAND, ENGLAND AND WALES.

By J. G. Kohl.   One Volume, 8vo. cloth....................$1.25

LIFE OF OLIVER GOLDSMITH.

By James Prior.   In One Volume, 8vo, boards...............2

OUR ARMY AT MONTEREY.

By T. B. Thorpe. 16mo, cloth.................... ....62 cts.
Paper covers.................. ....50 cts.

OUR ARMY ON THE RIO GRANDE.

By T. B. Thorpe. 16mo, cloth...................... ....62 cts.
Paper covers.................................. ....50 cts.

LIFE OF LORENZO DE MEDICI.

By William Roscoe.   In Two Volumes, 8vo, cloth..........$3

MISCELLANEOUS ESSAYS OF SIR WALTER SCOTT.

In Three Volumes, 12mo, cloth........................ ....$3.50
Half morocco................ ....$4.25

SERMON ON THE MOUNT.

Illuminated.   Boards..$1.50
  "   Silk $2.00
  "   Morocco super...$3.00

MISCELLANEOUS ESSAYS OF THE REV. SYDNEY SMITH.

In Three Volumes, 12mo, cloth........................ ....$3.50
Half morocco... ....$4.25

MRS. CAUDLE'S CURTAIN LECTURES............ ....12½ cts.

SERMONS BY THE REV. SYDNEY SMITH.
One Volume, 12mo, cloth.................................75 cts.

MISCELLANEOUS ESSAYS OF SIR JAMES STEPHEN.
One Volume, 12mo, cloth...........................$1.25

THREE HOURS; OR, THE VIGIL OF LOVE.
A Volume of Poems. By MRS. HALE. 18mo, boards.. 75 cts

TORLOGH O'BRIEN:
A Tale of the Wars of King James. 8vo, paper covers 12½ cts.
Illustrated..37½ cts.

AN AUTHOR'S MIND.
Edited by M. F. TUPPER.   One Volume, 16mo, cloth....62 cts.
Paper covers.......................................50 cts.

HISTORY OF THE ANGLO-SAXONS.
By SHARON TURNER.   Two Volumes, 8vo, cloth...........$4.50

PROSE WORKS OF N. PARKER WILLIS.
In One Volume, 8vo, 800 pp., cloth, gilt....................$3.00
Cloth extra, gilt edges......................................$3.50
Library sheep...$3.50
Turkey morocco backs.......................................$3.75
"   extra..$5.50

MISCELLANEOUS ESSAYS OF PROF. WILSON.
Three Volumes, 12mo, cloth.........................$3.50

WORD TO WOMAN.
By CAROLINE FRY.   12mo, cloth........................60 cts.

WYATT'S HISTORY OF THE KINGS OF FRANCE.
Illustrated by 72 Portraits. One Volume, 16mo, cloth...$1.00
Cloth, extra gilt..$1.25

9 780282 821289